FOLDED PAPER GEOMETRY

First published in Great Britain in 2025 by

LAURENCE KING

Laurence King
An imprint of Quercus
Carmelite House
50 Victoria Embankment
London EC4Y 0DZ

An Hachette UK company
The authorized representative in the EEA is Hachette Ireland,
8 Castlecourt Centre, Dublin 15, D15 XTP3, Ireland
(email: info@hbgi.ie)

A CIP catalogue record for this book is available
from the British Library

TPB ISBN 978-1-52944-048-5
Ebook ISBN 978-1-52944-049-2

10 9 8 7 6 5 4 3 2 1

Book and cover design by Alexandre Coco
Photography by Meidad Suchowolski
Videos by Yoram Ron
Project-managed and copy-edited by Rosanna Fairhead
Proofread by Patricia Burgess

Printed and bound in China by C&C Offset Printing Co., Ltd.

MIX
Paper | Supporting
responsible forestry
FSC® C104740
FSC
www.fsc.org

Papers used by Quercus are from well-managed
forests and other responsible sources.

FOLDED PAPER GEOMETRY

Essential Constructions for Designers

Paul Jackson

Laurence King Publishing

Introduction

1
Angles

2
Polygons

3
Divisions

4
Geometry from A4 Paper

5
Basic Solids

Introduction

Phrases such as digital age, digital studio and digital environment are well known among designers and makers. But what does *digital* actually mean?

Perhaps surprisingly, it comes from the Latin word *digitus*, meaning finger. So, digital age, digital studio and digital environment mean literally finger age, finger studio and finger environment. The word connects our most advanced technology with our simplest, most readily available tools: our hands. The contrast of means could not be more extreme.

So let's go digital. Let's make with our hands. What can old-school digital give us that new-school digital cannot?

Folding paper is as basic as old-school digital can get. Why? Because most making activities require us to hold a tool, but folding paper usually requires us to use nothing more than our hands. This

is a very unusual making practice, almost unique. It is as basic as basic comes. We are making something *by* ourselves, *with* ourselves, using no tools.

Most people reading this will have made a paper aeroplane, or perhaps worked from videos or books to make origami figures. These are 'models', of which there are many tens of thousands. Taking a step back, when the intention to represent is removed, we are left with one thing: folded geometry.

The geometry of paper folding is truly a glory. It is by turns simple, unexpected, elegant, audacious and, at its best, breathtakingly beautiful. Many constructions are design classics for the ages. Its aesthetics sidestep cultural values, reaching everyone, everywhere.

On one level, then, this book is a collection of poetic answers to prosaic geometric constructions. On another, it is a book of practical solutions, useful to every designer and maker. But perhaps most of all, it is a book that connects us with our instinct to transform something mundane – in this case, a sheet of ordinary paper – into something useful and beautiful, by means of a hand-based process that can be described as alchemic. In this way, the book raises questions for the thinking designer-maker to consider, about the role of technology versus hand skills, about the physicality of real-world geometry versus digital rendering, and about the connections between the designer, the method of designing and the designed object.

As you read through these pages, you are encouraged not just to look, but to make. Break out the copy paper and, instead of printing on it, fold it. Make a pentagon or a tetrahedron. Have you ever made and handled them before? Hold them, turn them around, examine them, connect them with others. What can you make? What do you learn? Even the simplest paper-folded two-dimensional polygons and three-dimensional forms are interesting, surprising and useful when you hold them in your hands and truly experience their physicality.

I hope this book persuades you that using your hands to make basic constructions is not only a great way to generate ideas for new designs, but also a deep and rewarding experience in itself. Your ten digital designers are the best design tools you will ever have, and the geometry they can make is the most useful of all visual languages.

Paul Jackson

Folding as Geometry

A sheet of paper is finite. It has edges, corners and a plane surface, all of which can be measured precisely for length, angle and area. This means that the position of a fold line made on a sheet can be plotted using X and Y coordinates. It follows, then, that because the length and orientation of every fold can be described, a crease pattern made on a sheet of paper is an inherently geometric construction (see right).

A fold itself can be described as the perpendicular bisector of two distinct points (these points may or may not be on the sheet) or, conversely, a unique line that connects two distinct points (see right, below).

The entirety of the technique and art of paper folding derives from these two axioms (self-evident truths). They are simultaneously simple and infinitely variable. They can be expanded to a set of seven axioms known as the Huzita-Justin-Hatori axioms, which describe the complete set of possible operations to make a precisely located fold on a sheet of paper. If you have an interest in the mathematical foundations of folding, they are well worth reviewing.

If every fold is inherently geometric, the challenge is to assemble a collection of folds into a crease pattern that has some added interest, meaning or beauty. The process is somewhat analogous to sound and music: there are sounds around us all the time, but music is the organization of sound into a meaningful structure.

The organization into meaningful constructions of the geometry inherent in the action of folding is the subject of this book.

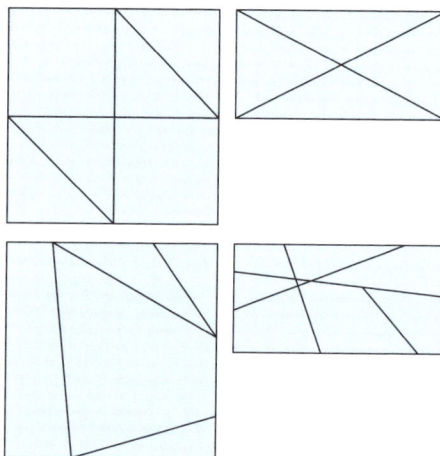

The top row shows creases that have been geometrically aligned and which are simple to describe. The second row shows creases placed at random. Nevertheless, their positions can all be precisely described.

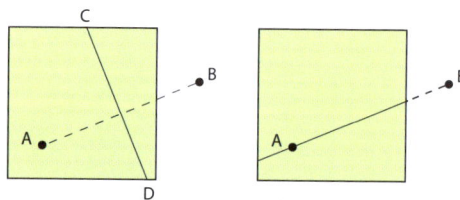

Crease CD is formed when point A is placed on top of point B. The crease is the perpendicular bisector of the line connecting A and B. Although the fold is unique, there is an infinite number of positions for A and B. Any two points equidistant from the fold and perpendicular to it will describe the position of the fold.

A unique fold can also be made that connects A and B.

How to Use This Book

This is a practical book, a book of things to make that you should find useful as a designer-maker. But don't just read the text! Don't just look at the illustrations! You should *make* as many of the projects as possible.

Only by making them, by turning them around and around in your hands, by playing with them, by studying them as physical objects, will you understand their relevance to your creative practice. Reading the book will give you knowledge, but making the examples will give you a true understanding. For this reason, you are strongly encouraged to break out the copy paper ... and make, make, make.

Some of the projects will be time-consuming to make and/or assemble. That's a good thing. Making slows the brain, allowing you time to absorb and assess each intermediary shape or form. For sure, the destination is important, but the steps along the journey of achievement should not be discounted or overlooked. Remember: each step is itself an example of folded geometry. You are here to learn, and as much can be learned from the process of folding as from the result, if not more. Folding should be a pleasure, not a chore.

In short, the more projects you make, the more you will learn from this book.

Materials and Equipment

Paper and card

Many of the constructions in the book can be made very well using copy paper, the kind used in office photocopiers and printers. Most of the world uses copy paper known as 'A4', which has the dimensions 297 × 210mm and a weight of 80 grams per square metre (80gsm). In North America, the common paper is American Letter Size (ALS), which has the dimensions 8½ × 11in and a weight of 20lb per 2,000 sheets.

Of course, other papers may be used, but generally, there is no need to use anything more exotic than copy paper. It is easy to find, inexpensive, excellent to fold and aesthetic. You may wish to increase the visual appeal of what you make by using coloured copy paper rather than white.

Later in the book, some of the 3-D constructions should be made with heavier weights of paper, often described as 'card stock' or 'Bristol'. The exact weight of the card is unimportant, but anything around 200gsm or 54lb is ideal.

Many geometricians consider the term 'corner' imprecise, preferring the more accurate 'vertex' or its plural, 'vertices'. However, since we would say 'the corner of the table' to describe a made object and since this book is about making objects from paper, it would seem acceptable to use the word 'corner' in this context. Accordingly, it is used throughout the book.

Equipment

Many of the projects in this book are made only by folding, but some will nevertheless require the paper or card to be cut to size beforehand or trimmed to shape at the end. Some of the 3-D constructions will need a little glue. For these procedures, you will need:

- Self-healing cutting mat

- Sharp craft knife

- 30cm/1ft ruler

- Hard pencil and eraser

- Paper glue (stick or tube)

Your Working Environment

Rather than opening the book and simply beginning to work from its pages, it is recommended that you take a little time to prepare your workspace.

- Tidy up. Clear away the clutter to make a clean and organized space.

- Make sure you have all the materials and equipment you need before you begin.

- Turn off your phone, or at least put it where you can't see it.

- Move your table to a place where the light is frontal (such as from a window), not coming from above. This will create strong shadows across your paper so that you can see what you are folding.

- Work methodically. If you become stuck, take a break and try again.

A well-organized space will increase your pleasure, your work rate and your sense of achievement. When you have finished, attach the models you have made to a pinboard and/ or show them to colleagues and friends. People of all ages are generally intrigued that so much can be achieved using a simple piece of paper, and will want to follow your progress through the book. Share what you have made with them. Share also your satisfaction and your thoughts. Teach them your favourites.

Key to Illustrations

Valley fold

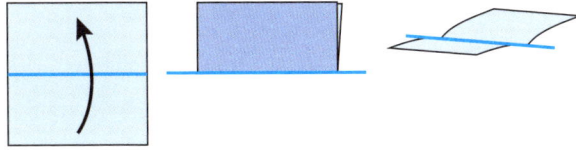

Mountain fold

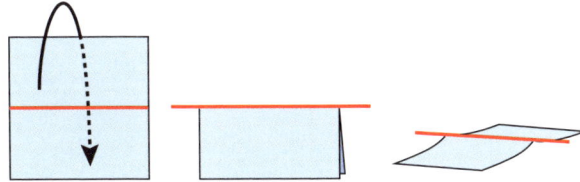

Cut

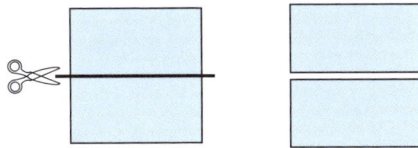

Glue here

Align the dots

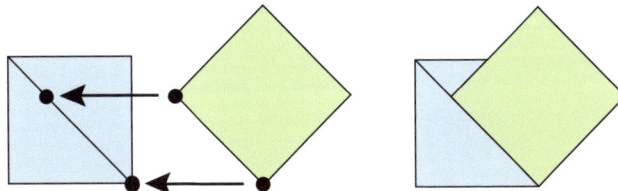

Important
alignments

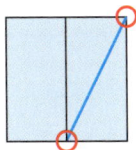

Angles

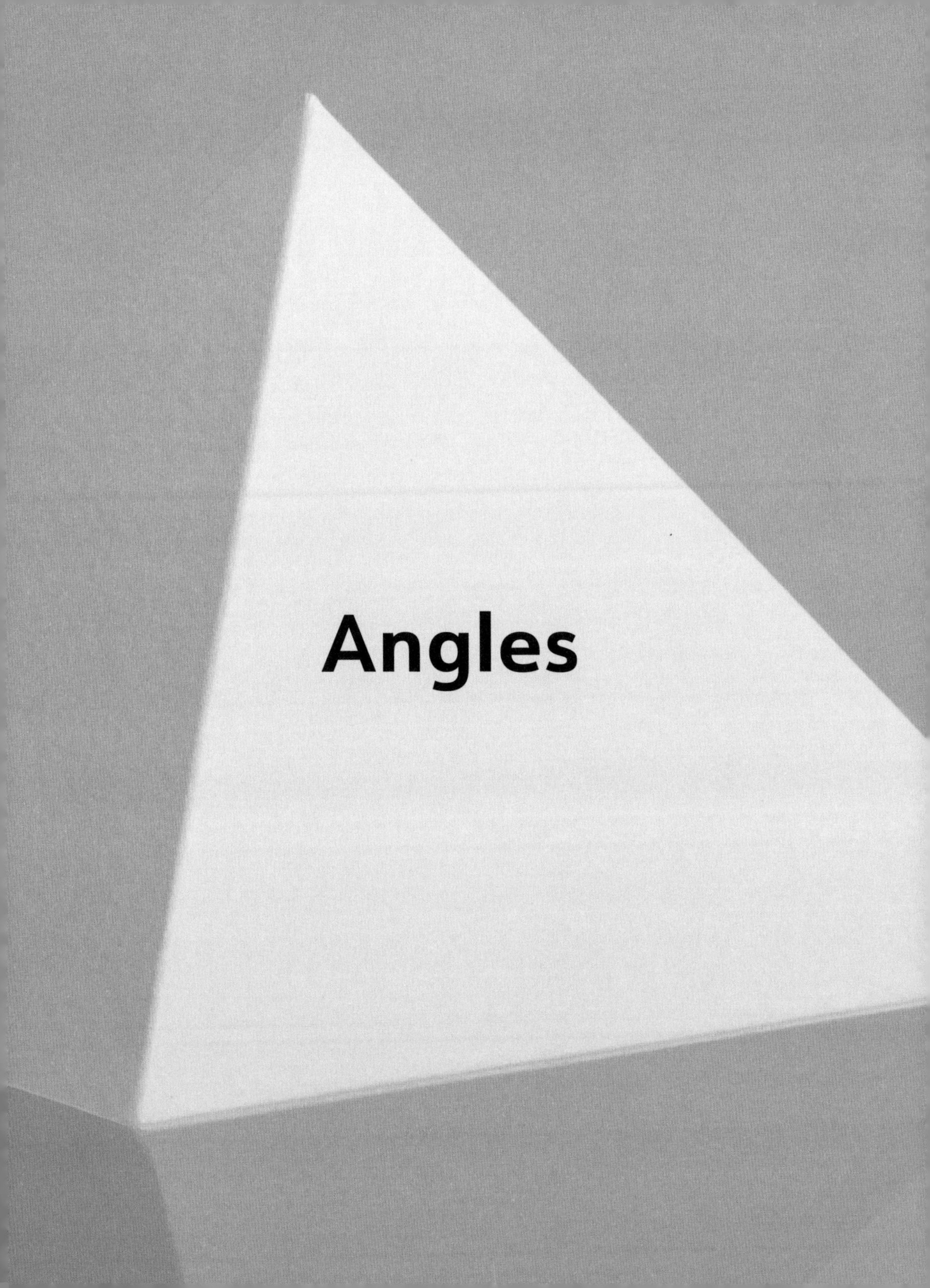

1
Angles

When we fold paper, angles are the inevitable consequence of terminating a fold at an edge or a corner. Of course, there are a great many possible angles, but only a few are generally useful. It is these angles that we need to know how to construct.

This chapter will introduce simple methods of making these useful angles. Some of these methods are intuitive and you may already know them, whereas others will be new and more subtle. Learning them all will give you a great command in constructing all manner of polygons and three-dimensional geometric figures.

Just one or two methods are presented for constructing each angle. However, they can all be constructed in several ways, which you may wish to explore for yourself.

1.1   90°

All manufactured papers come with ready-made 90° corners.
Sometimes, however, the corners or edges are torn or crumpled
and we need to redefine the perimeter to create a pristine sheet.
This can be time-consuming using geometry equipment, such as
a protractor, but the task becomes quick, easy and accurate if the
paper is folded.

1

Here is the damaged sheet.
Two edges are rough, but
this method will also work
with one, three or four
rough edges.

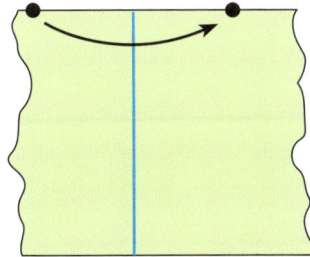

2

Double a straight edge back
on itself. If all four edges are
rough, simply make a vertical
fold as shown, and continue
through the steps.

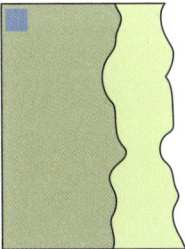

3

This immediately creates a 90°
corner. Perhaps this is all you
need to do to create a usable
sheet, but we will continue.

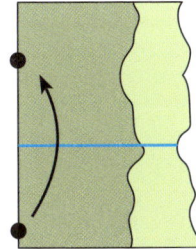

4

Fold the folded edge
back on itself.

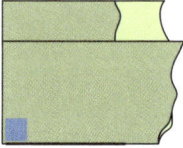

5

This creates a second 90° corner, now through four layers of paper.

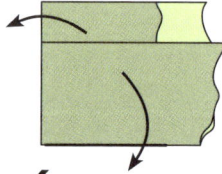

6

Unfold the paper.

7

When two crease lines cross, if one angle at the intersection is known to be 90°, all the others must also be 90°. This method can be used several times on a sheet of paper, depending on the amount of peripheral damage to be removed.

When you have created the perimeter you need, use the unfolded creases as a guide and trim off the excess paper.

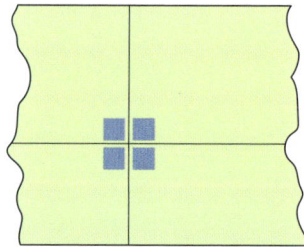

When making the folds, try to double back as much of the top-layer edge as possible over the edge underneath. This will increase the accuracy of your 90° corner.

Angles

1.2 45°

If you have ever folded paper, very probably the first angle you made that was not 90° was 45°. It is the most useful and intuitive of all angles to fold.

Two methods are presented here. The first is the traditional way, while the second is a subtler, rather beautiful method, showing that there are many ways of achieving even the simplest result.

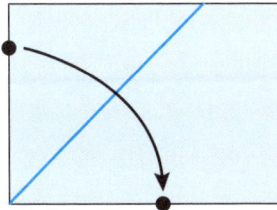

1

Take any rectangle of paper and fold a shorter edge to lie on top of an adjacent longer edge, running the fold into the common corner.

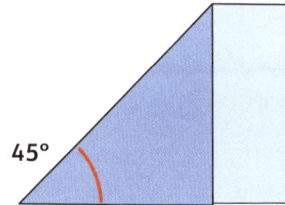

2

Since the original angle of the corner was 90° and the paper now has a double thickness, the new angle must be half of 90°, which is 45°.

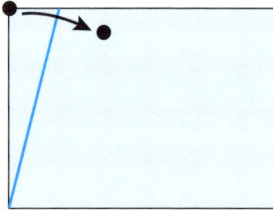

3

There is another way to create 45°. Fold in an edge at a random angle. Anything at all!

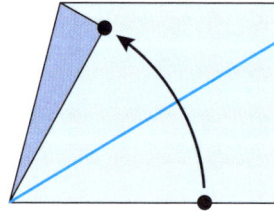

4

Fold in the adjacent edge, allowing the raw edges of the paper to butt against each other exactly.

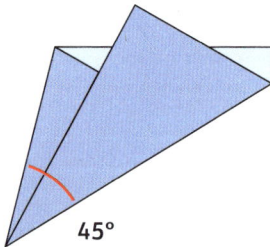

45°

5

The angle is 45°. Why? Because, as in the first example, all the paper is now a double thickness, so the angle must be half of the single-thickness 90°.

In this way, the two parts of the angle can be large or small, or even equal, depending on how the two edges are folded. This will still create a wholly useful angle, but it will add significant visual appeal. It also means that the external edges are folded (closed) so that nothing can slip inside, between the layers.

1.3   30-60-120°

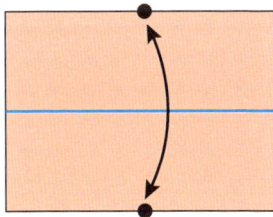

This construction is perhaps the most poetic in all folded geometry. It is intuitive and very simple to create a 45° corner from a 90° corner because the angle is simply folded in half. However, angles of 30° and 60° involve dividing 90° into three, which intuition says must be impossible, or at least awkward. That is not so; the solution is simple and breathtakingly beautiful. Prepare to be awed.

1

Begin with a rectangle. Create a centre fold on the longer axis. Unfold.

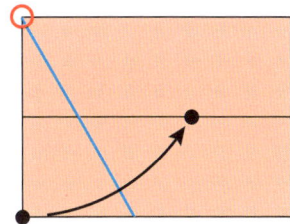

2

Very carefully, fold the bottom corner upwards to lie on the centre line, such that the fold you make runs exactly to the top corner. Take your time and be accurate.

60°

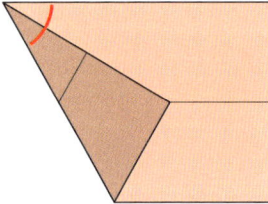

3

Astonishingly, this simple location fold has created an angle of 60° at the top corner. There are many proofs of this, some requiring just common sense and no knowledge of maths or geometry. Can you prove why the angle is 60°?

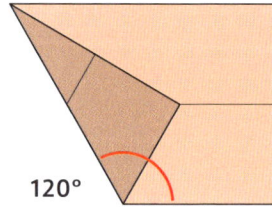

120°

4

If the top angle is 60°, it follows that the bottom angle is 120°, or twice 60°. This is because the perimeter of the paper rectangle already has angles of 90°, 90° and 60°, so to total 360° (the number of degrees in any quadrilateral), the fourth corner must have an angle of 120°.

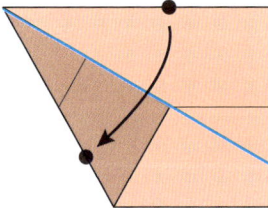

5

Bring the top edge down to the folded edge.

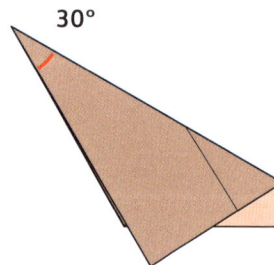

30°

6

Folding the corner in half must halve the 60° angle to 30°. Note that the fold runs along the edge separating the double layer from the single layer.

1.4   135° and 67.5°

These angles relate specifically to the geometry of octagons (eight-sided polygons). 135° is the angle inside a regular octagon at each of the corners, and 67.5° (half of 135°) is the angle from a corner to the centre point of the figure.

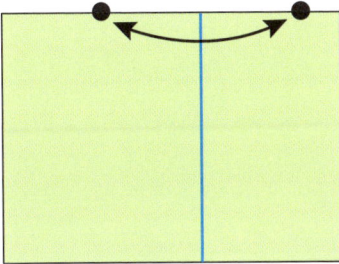

1

Take a rectangle of paper and fold the top edge back on itself. The exact position of this fold is not critical. Unfold the crease.

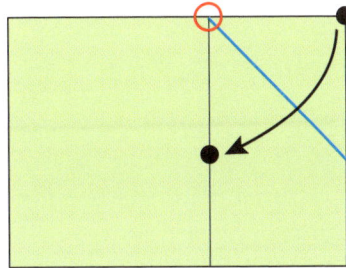

2

Take one of the top corners and fold it to the crease line.

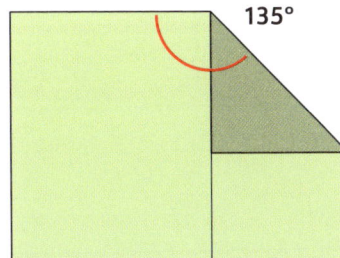

3

The angle created is 135°, that is: an angle of 90° plus an angle of 45°.

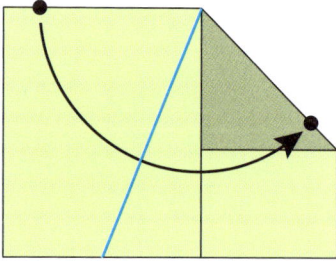

4

Fold down the top raw edge
of the paper to lie along the
folded edge, folding the 135°
corner in half.

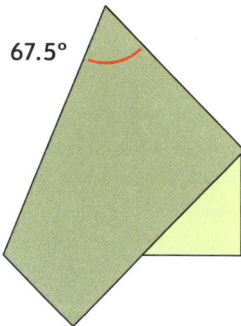

67.5°

5

The result is an angle of 67.5°.

This simple construction, like
many in the book, is quick,
accurate and pleasing to
make. To make these angles
using geometry equipment
would be slower and would
involve some numerical
calculation, thus introducing
the possibility of error. The
paper-folding method is
clearly advantageous.

1.5 Folded Protractor

scan for video

The preceding examples have shown how to create specific angles, so here, by contrast, is a folded protractor that shows how to make at least nine different angles and, as a bonus, two common types of triangle.

Make it as large as possible so that when you use it to measure angles, the large scale will help you to mark the angles accurately. A small paper protractor will lack the necessary accuracy.

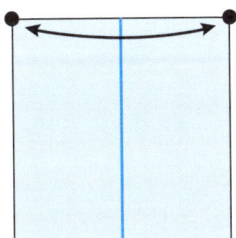

1

Begin with a square of paper. If you have only rectangles to hand, use one of the methods described in Chapter 3 or Chapter 5 to make a square. Fold a vertical centre line. Unfold the paper.

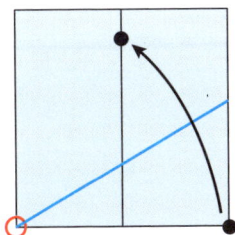

2

Fold dot to dot, bringing the bottom corner to lie on top of the crease, a little below the top edge of the paper. The new fold will run exactly to the other corner at the bottom of the square.

Note that this is the same folded step seen previously in this chapter to make a 60° angle (see 1.3).

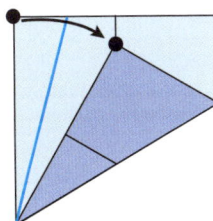

3

Fold dot to dot, allowing the two raw edges of the square to butt against each other.

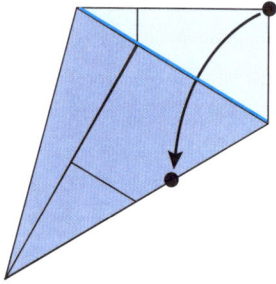

4

Fold dot to dot, bringing the corner down to the folded edge.

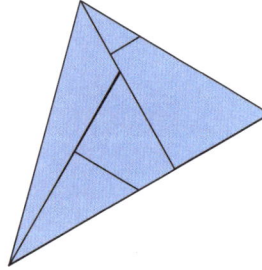

5

The folding is now complete and we can begin to identify the many angles.

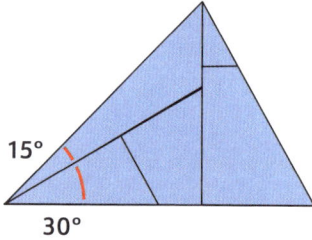

15°

30°

6

The two angles in the corner are 15° and 30°. They are associated with the 60° geometry of triangles, but in this instance, when added together ...

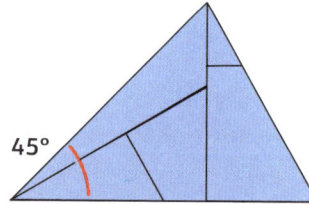

45°

7

... the two angles total 45°, an angle associated with squares and rectangles.

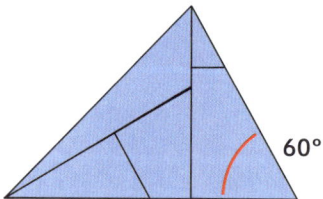

60°

8

The other corner at the bottom of the triangle has an angle of 60°. Can you see why this is so?

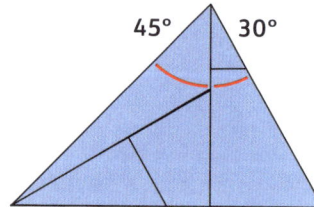

45°   30°

9

The two small angles at the top of the protractor are 45° and 30°. When added together ...

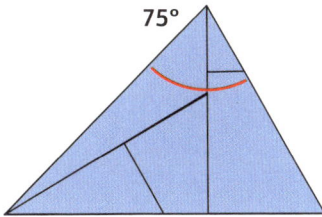

75°

10

... they create an angle of 75°.

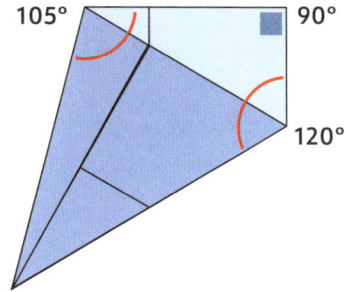

105°   90°

120°

11

By unfolding the protractor to the shape in step 4, three more angles can be found: 90°, 105° and 12°.

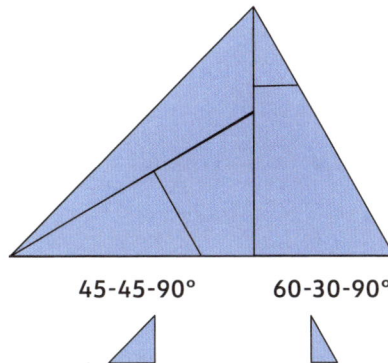

45-45-90°   60-30-90°

12

Additionally, two common triangles can be found, with angles of 45°, 45° and 90°, and 60°, 30° and 90°.

The angles found divide 90° into halves (45°) and thirds (30° and 60°). We have also found every 15° increment from 0° to 120° (15°, 30°, 45°, 60°, 75°, 90°, 105° and 120°).

If you doubt that some of the angles are what they are claimed to be, analysing the construction of the protractor is a fun exercise in geometric proof. Can you find any other angles? Can you create a protractor that would divide 90° into quarters, with angles of 22.5°, 45°, 67.5° and 90°?

②

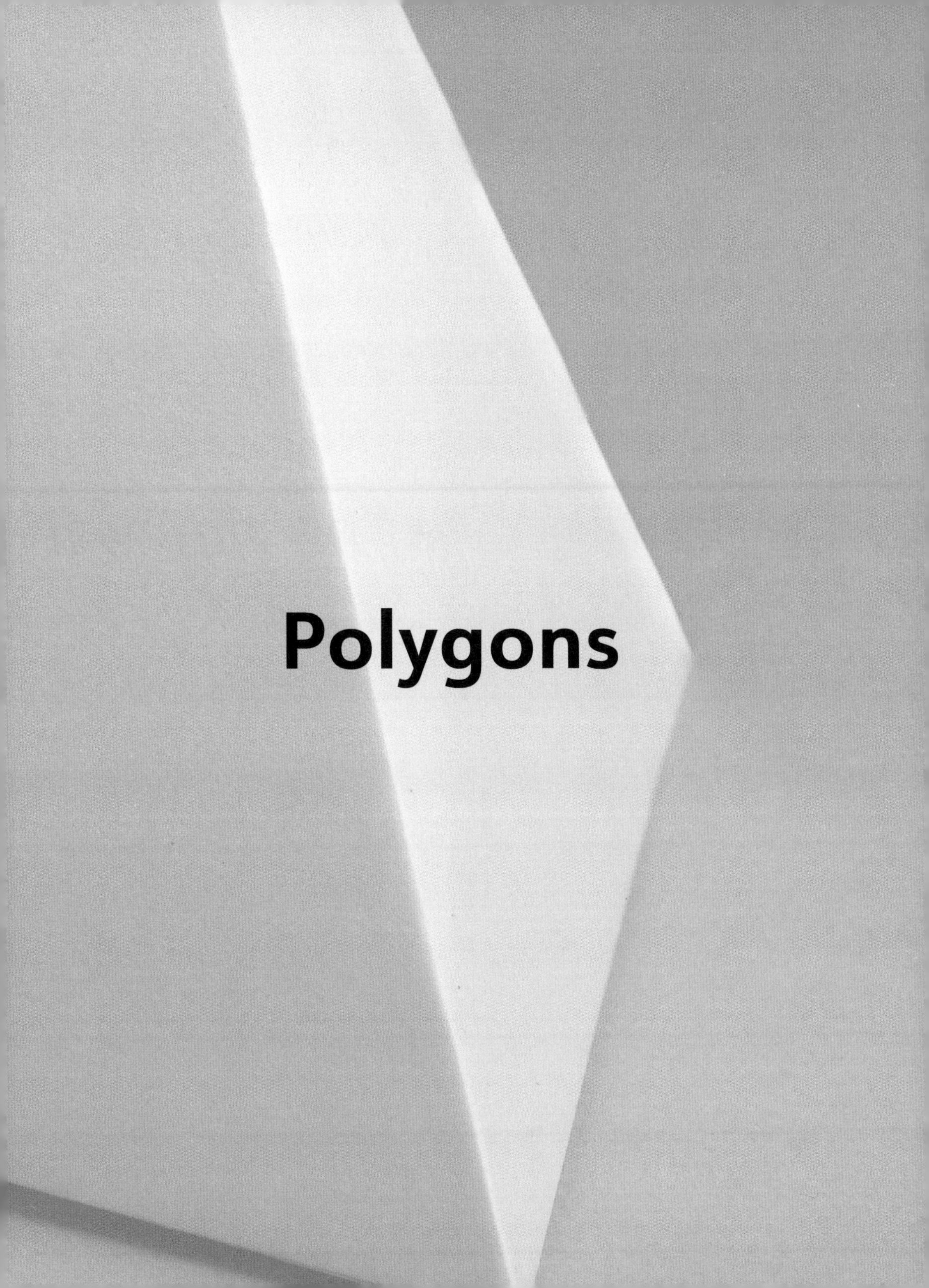

Polygons

2
Polygons

The angles we made in Chapter 1 connect to create polygons
(flat, enclosed geometric shapes bounded by straight edges).
There is an infinite number of possible polygons, but only
a few are generally useful. This chapter focuses on those few,
giving simple methods of constructing them by folding paper.
Many of these methods use elegant folding sequences that
exploit angles and relative proportions with great ingenuity.
They are in themselves design classics, poetic ways of
achieving prosaic constructions.

You should also refer to Chapter 4, which shows how A4
paper – commonly available outside North America – can
be used to create a number of polygons using the specific
proportions of the sheet.

2.1   Equilateral Triangle from a Rectangle

An equilateral triangle has three equal sides and three equal angles of 60°. It is perhaps the simplest of all polygons, having the fewest possible sides and three equal angles. The key to making this figure is to know how to fold an angle of 60°, as explained in 1.3.

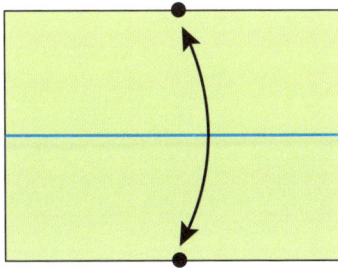

1

Use any rectangle, such as A4 or 8½ × 11in. Fold the long sides together to create a centre fold on the long axis. Unfold.

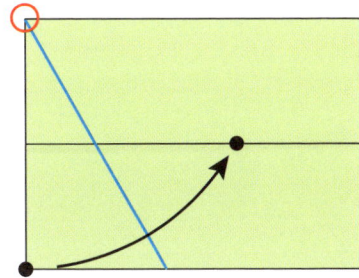

2

Very carefully, fold the bottom corner upwards to lie on the centre line, such that the fold you make runs exactly to the top corner. Take your time and be accurate.

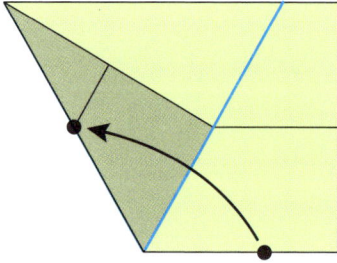

3

Fold the remainder of the
bottom edge upwards so that
it lies along the folded edge
you have just created. Fold the
paper flat.

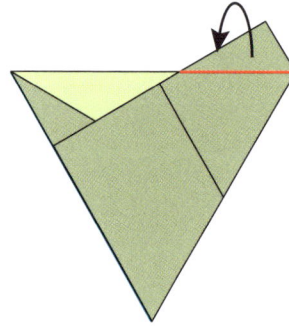

4

Fold the excess triangle at the
top out of sight. Depending on
the shape of rectangle you are
using, the size of this triangle
will vary.

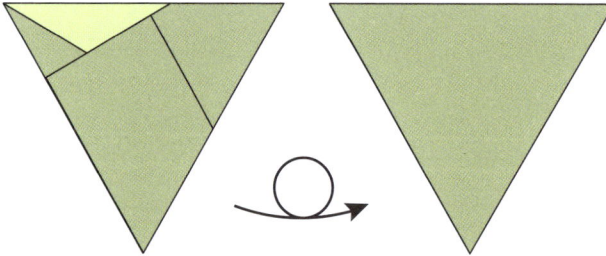

5

Turn the paper over.

6

Complete. The polygon is
an exact equilateral triangle.
If you wish to create just
a simple, one-layer triangle,
open all the folds and trim
off the excess paper around
the triangle.

Note that in step 4, the corner
at the bottom of the triangle
has been made by exactly
trisecting the bottom edge
of the paper rectangle; that
is, the edge has been folded
into three equal angles, each
of 60°. This complex trisection
was achieved with great
economy of means.

2.2 Equilateral Triangle from a Square

scan for video

The method of folding a 60° angle can be used in a different way to create an equilateral triangle from a square of paper. If you like geometric puzzles, you might want to find a proof for this variation.

If you have only rectangles of paper to hand, this chapter and Chapter 4 give several different methods of creating squares from rectangles. Use one of them to create a square, then return to this page.

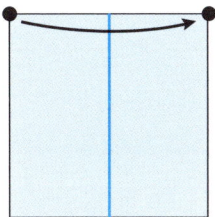

1

Fold a square in half from side to side.

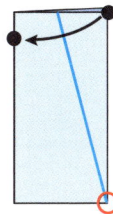

2

Note that the folded edge is at the left. Fold in one of the top corners to touch the fold, such that the fold you are making runs exactly down to the bottom corner. This fold can be a little awkward to make, so take your time to ensure accuracy. With a little practice, it becomes easy to do.

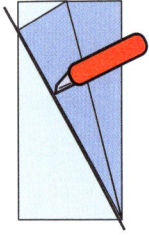

3

Use a ruler and a knife to trim off the excess paper. The cut will be accurate if you place the ruler on the excess paper, not on the triangle. This is because when on the excess paper, the ruler will hold everything flat to ensure an accurate result.

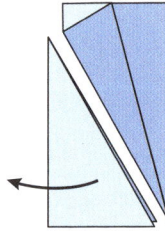

4

The triangle is separated from the excess paper. Open it.

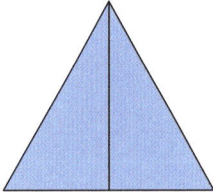

5

Complete. The equilateral triangle will be accurate if the fold in step 2 was placed accurately.

Note how large the triangle is in the square. Surprisingly little paper is discarded. This lack of waste is very pleasing, indicating an extremely efficient method. A very slightly larger equilateral triangle can be constructed inside a square if the triangle is tilted a little, but the construction is complex and perhaps not worth the effort, given the minimal benefits.

2.3   120-30-30° Isosceles Triangle

Several triangles can be folded using the technique for creating an angle of 60°, but which don't necessarily use 60° itself. One of them is the triangle shown here. 'Isosceles' means that the triangle has two sides of equal length and two equal angles. In this example, the triangle is not created by trimming away the excess paper; rather, it is multilayered and self-locking.

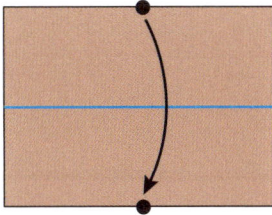

1

Use any rectangle, such as A4 or 8½ × 11in. Fold the long sides together to create a centre fold on the long axis.

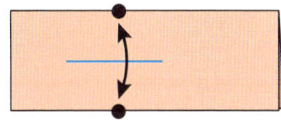

2

Note that the fold runs across the top of the paper. Pick up the top layer and fold dot to dot, folding the paper in half. This need not be done across the whole length of the paper, as shown. Unfold.

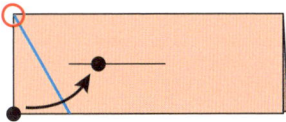

3

Pick up both corners at the bottom (one is behind the other) and lay them on the crease in the centre. Do this in such a way that the new fold runs exactly up to the top corner. If you have made other examples in the book, you will recognize this as a way to create an angle of 60° (here at the top left corner).

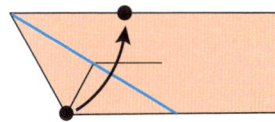

4

Fold dot to dot, folding the 60° corner in half. Can you deduce what the other angles will be?

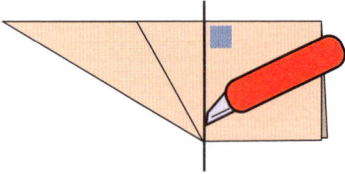

5

Trim off the excess rectangle of paper on the right, under the knife.

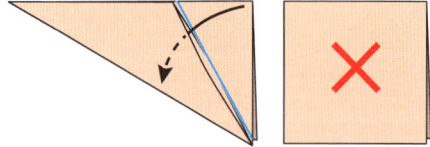

6

The rectangle can be discarded. Fold the corner at the top right into the pocket. This will lock the paper flat. Be sure to tuck it deeply and cleanly inside.

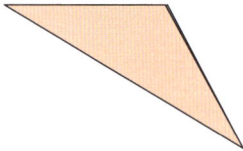

7

This is the completed triangle.

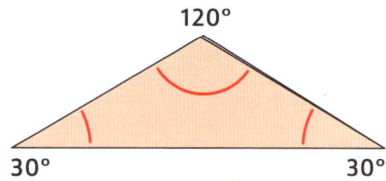

120°

30°   30°

8

The large angle is 120° and each small angle is 30°. The advantage of this locked, multilayer example is that it is sturdier than a single-layer example, while retaining accuracy. The disadvantage is that it is smaller.

2.4 60-30-90° Triangle

Here's another in the family of triangles that uses the method to create a 60° angle, and another that is multilayered and self-locking. It has the same angles as the traditional 60-30-90 set square and can be used to perform many of the same constructions.

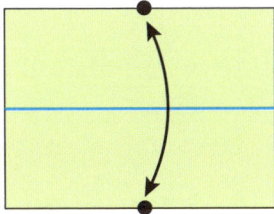

1

Use any rectangle, such as A4 or 8½ × 11in. Fold the long sides together to create a centre fold on the long axis. Unfold.

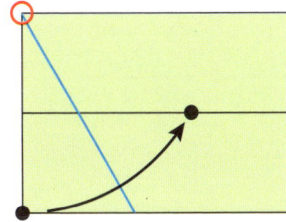

2

Very carefully, fold the bottom corner upwards to lie on the centre line, such that the fold you make runs exactly to the top corner. Take your time and be accurate.

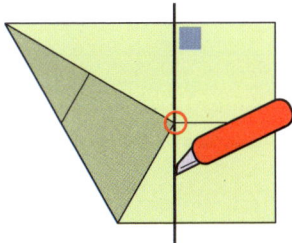

3

Trim the rectangle with a cut that touches the corner of the triangle. Make sure the cut is at 90° to the top and bottom edges.

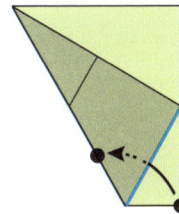

4

Fold dot to dot, as shown. This will trisect the corner at the bottom.

Important: Tuck the triangle *under* the top layer. This will create a pocket to be used in step 6.

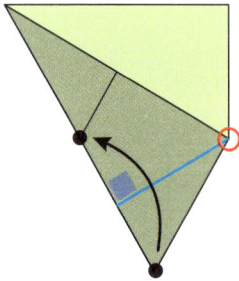

5

Again, fold dot to dot, as shown. Note the location points of the fold.

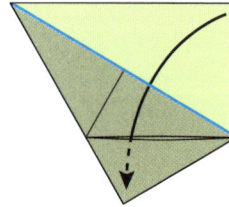

6

Tuck the corner of the paper rectangle deep into the pocket. It should fit exactly inside.

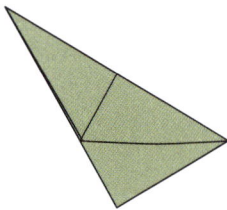

7

Press the edges flat with some force to ensure a crisp, flat triangle. Turn it over.

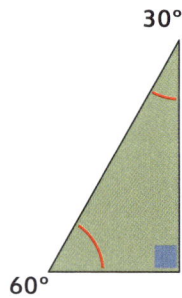

30°

60°

8

Complete. If it is not sitting completely flat, continue to crease the edges with force. The paper will eventually be ironed sufficiently flat to sit well on a surface.

2.5   45-45-90° Isosceles Triangle

This is surely the simplest triangle to fold because it uses the natural 90° geometry of a square or rectangle. The version here is multilayered and self-locking. The lock itself is unusual and was discovered only after many hours of creative experiments to find a way of making a large, self-locked 45-45-90° triangle. There are many small examples, but this is perhaps the largest.

I once taught this design at an international origami convention, not as an example of folded geometry, but as an origami model. Expecting a showy result, my class were somewhat confused. I told them that it was called 'Every Home Should Have One', explaining that it was the deliberate antithesis of a good origami design, being flat, plain and a simple, dull-looking 45-45-90° triangle. Hence the ironic title. Later, quite a few of my students told me it was their favourite class of the convention. 'It made me think,' was the common comment. Moral: it's all about the context.

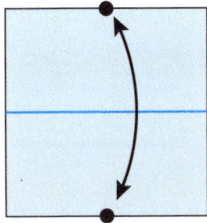

1

Fold a square in half. Unfold.

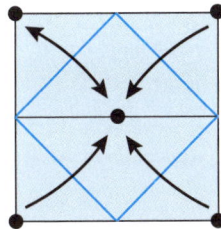

2

Fold each of the four corners to the centre line. Three can remain folded, but unfold one corner.

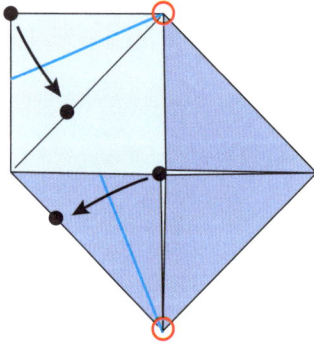

3

Before folding, look carefully at the illustration to understand where the two new folds will be placed. The upper fold moves the top edge of the paper square to lie along the unfolded crease. The lower fold moves the corner of the triangle to the outer edge.

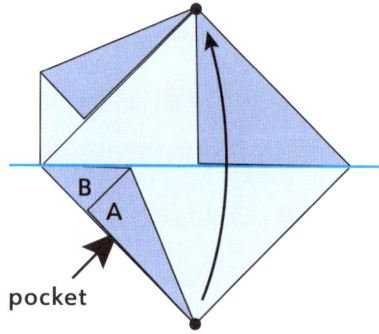

4

Check that your paper matches the illustration. Identify the pocket between layers A and B. Now fold the bottom corner up to the top corner along the existing crease.

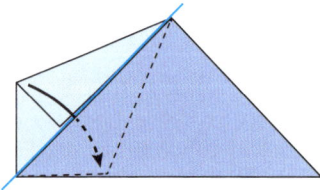

pocket

B A

5

Tuck the triangle into the pocket identified in the previous step.

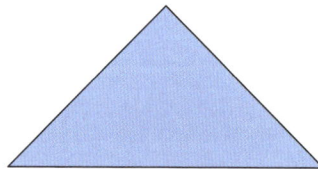

6

Complete. Smooth the edges very flat. The result is well locked and pleasingly clean on both sides, and the triangle will lie very flat on a surface.

2.6   Uncreased Square from a Rectangle

Almost all bought papers are not square, but rectangular.
To make a square, they must be trimmed. There are times
when it is necessary for the design to make a square without
a crease across it – unwanted creases can be very unsightly.
Here is a method of making a clean, blank, uncreased square.

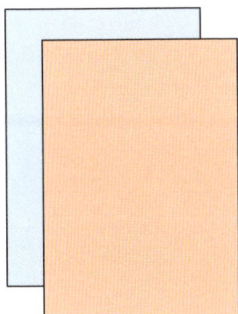

1

You will need two identical
rectangles, here shown in
different colours, for clarity.

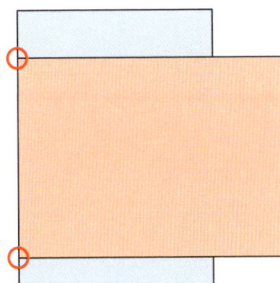

2

Turn one of the rectangles
sideways. Align the corners of
the sideways rectangle exactly
on to the long edge of the
vertical rectangle, as shown.
Take your time.

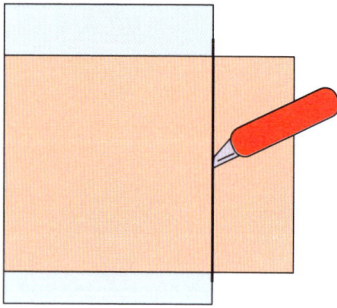

3

Cut the sideways rectangle along the edge of the vertical rectangle. This can be done by marking the point with a pencil, by making a fold and cutting along the fold, or simply by cutting with no preparation. There is no best way, simply personal preference. The important thing is to choose a method that, for you, guarantees accuracy.

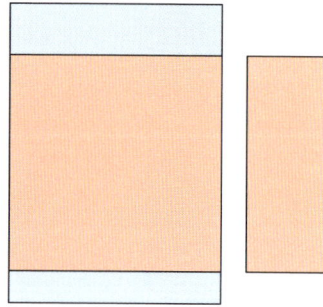

4

The offcut can be discarded.

5

Complete. With no reference points on the square to confirm the accuracy of its construction, the blank square may seem inaccurate. However, with practice and perhaps a little experimentation, preparing a square this way becomes quick and very reliable. Be careful, though: there is nothing more frustrating than trying to fold an almost-square square.

2.7 Creased Square from a Rectangle

If you know a way of making a square from a rectangle – any rectangle – this is probably it. It's not a great way because it leaves a crease that you may not want, but it's easy and reliable. To my surprise, it's not known to all my students; there are always a few (maybe 5 per cent) who have never seen it. In a book about the basics of folded geometry, it's a must-have.

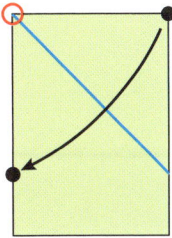

1

Take any rectangle and bring one corner across to the opposite long side. Flatten the short side exactly on top of the long side, running a crease exactly into the corner, as shown.

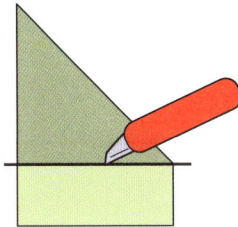

2

Trim off the excess paper.

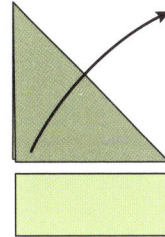

3

Open the triangle.

4

Complete. This is the classic way to make a paper square. The acceptability of the crease may depend on whether or not it is used in the final design.

Instead of trimming off the excess rectangle with a knife, here's an alternative method using no equipment.

Turn the paper over and make a fold across the paper where the cut would be made.

Use the edge of the triangle beneath as a guide to place the fold. Fold the crease backwards and forwards a few times to weaken it, running your fingernail along it for extra pressure. Then run your tongue along the folded edge, turn the fold the other way and repeat. This small amount of moisture will significantly loosen the fibres and the rectangle will

tear off with surprising ease. This – admittedly unhygienic – method should not be used if the paper is old or has been printed on. Paper fresh from the packet is best.

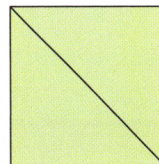

2.8   How to Trim 8½ × 11in Paper to A4

The 8½ × 11in paper commonly used across North America –
usually known as American Letter Size – is of little use to those
with an interest in folded geometry. Much more useful is the A4
paper used in the rest of the world, the two sides of which are in
the ratio 1:$\sqrt{2}$. It is possible to trim Letter paper to this ratio, but its
dimensions will be a little smaller than A4. This should not matter
since we are interested in geometry, not scale.

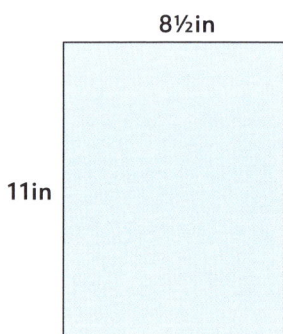

8½in

11in

1

This is 8½ × 11in paper. It's
a little wider and a little
squatter than A4 paper.

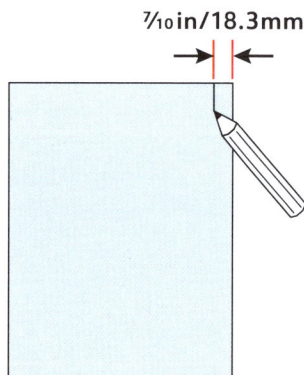

$^{7}/_{10}$in/18.3mm

2

Mark a point $^{7}/_{10}$in (or ¾in,
depending on how your ruler
is calibrated) or 18.3mm along
the top edge. Repeat along
the bottom edge.

3

Connect the two marks with
a ruler and trim the paper,
as shown.

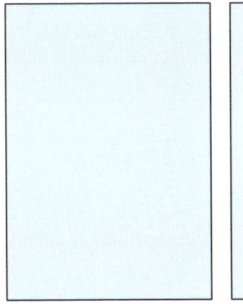

4

Discard the trimmed edge.

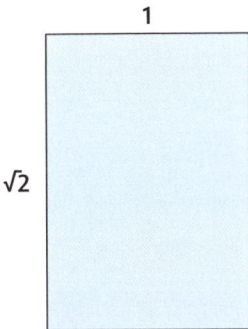

1

$\sqrt{2}$

5

The new sheet is proportioned
$1{:}\sqrt{2}$ and may be used to
create any of the constructions
in the book that use A4 paper.

The preference for 8½ × 11in paper
in North America has long been a
minor irritant. It means, for example,
that copier machines have a slider in
the paper tray to accommodate the
different widths of 8½ × 11in and A4
paper. Also, envelopes, folders and
binders are proportioned differently.
Further, the weight of this paper (and
of other papers used in North America)
is measured in pounds, not in grams,
frequently leading to confusion. A single
paper standard implemented around
the world would be preferable, but it
seems unlikely, at least for the present.

2.9 Pentagon from a Square

scan for video

Whereas the geometry of a regular triangle (an equilateral triangle) is compatible with a square, the geometry of a square is barely compatible with that of a regular pentagon – 'regular' meaning that all the sides and angles are equal. This relatively simple regular polygon with just five sides *can* be folded from a square of paper, but the method is very forced and it is not a thing of beauty. In any case, a long sequence of contrived folds will inevitably become inaccurate, so the smart paper folder would prefer to use a much simpler folding method that is only minimally inaccurate. The results will be the same.

There are several elegant methods of creating almost regular pentagons containing only the smallest of errors. The method described here is attributed to Fred Rohm (US), one of the pioneers of creative origami in the West in the 1950s and 60s. The star of the sequence is the simple fold in step 3, from which the geometry of a pentagon can be developed.

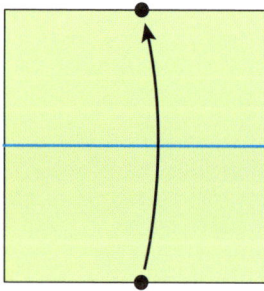

1

Fold the bottom edge up to the top edge and crease flat.

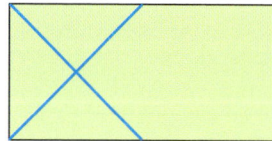

2

At the left, make two folds in the shape of a large X. The X itself is not of interest; we are interested only in the point where the two folds intersect.

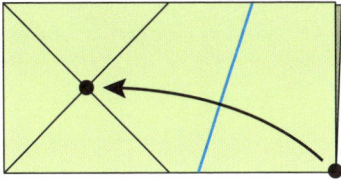

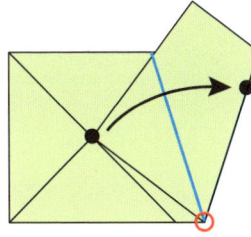

3

Before folding this step, check that the fold you made in step 1 runs along the bottom of the paper, not along the top. Bring the bottom corner on the side of the paper away from the X, to the intersection of the two X folds. Make a fold. This unlikely step creates the angles necessary to make a regular pentagon with only minor errors.

4

Fold the edge back to lie on top of the outside edge.

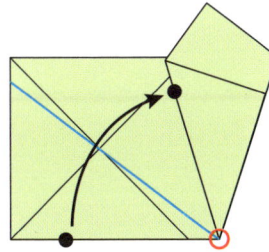

5

Fold the bottom edge upwards to butt against the folded edge. Try hard to keep the bottom corner clean and tidy.

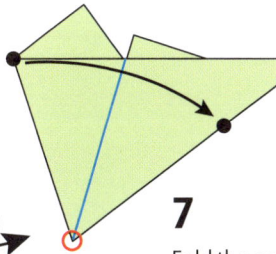

6

Turn over.

7

Fold the corner at the bottom in half.

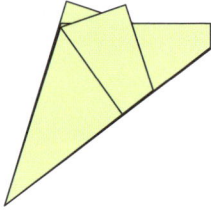

8

This is the result. There should be three edges on both the left-hand side and the right-hand side that line up perfectly with one another. If the zigzag of layers is unfolded back to step 3, you will see that the bottom corner is divided into five equal angles. This equality means that the angles required to create a regular pentagon have been established. The sequence to this point is a remarkable piece of folded geometry.

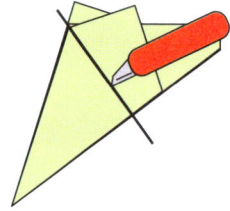

9

Use the edge closest to the sharp corner at the bottom of the paper as a guide to make a cut. Position the ruler with care to cut exactly along the edge.

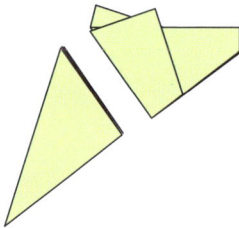

10

Discard the upper portion of the paper. Before doing so, take time to unfold it and lay it out flat. You will notice that the pentagonal hole occupies a pleasingly large area of the paper square. The folding method is not only beautiful, but also very efficient.

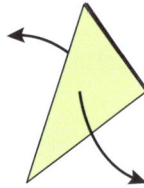

11

Unfold the zigzag.

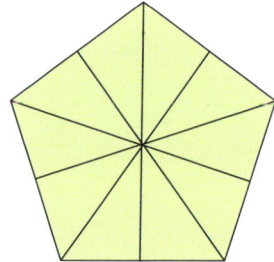

12

Complete. Folded and cut with care, the pentagon will be indistinguishable from a regular pentagon. Purists may baulk at its inaccurate structure, but the errors are so small that, for practical real-world purposes, they can be ignored.

Polygons

2.10   Hexagon from a Rectangle

The geometry of a regular hexagon is similar to that of an equilateral triangle, so it is not surprising to learn that the method of constructing it uses the method to create a 60° angle.

1

Use any rectangle. Fold in half down the long axis, then fold the long quarters.

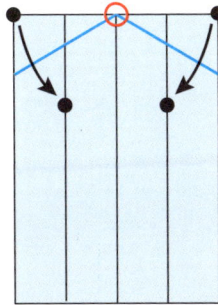

2

Fold in both top corners to lie along a quarter fold. Note that both folds meet exactly at the midpoint of the top edge. Make this new corner as neat as possible. This technique is the same as the one used in 1.3 to create a 60° angle. The difference is that here, two 60° angles are placed back to back to create one angle of 120°, the angle at each of the six corners of a regular hexagon.

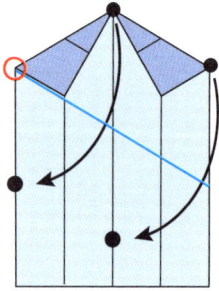

3

Fold the top corner down to lie along the side edge of the rectangle. The crease will run exactly into the corner, as shown. Note how the corner at the top right will fold down to lie on top of the centre fold. This second alignment will check the accuracy of your folding.

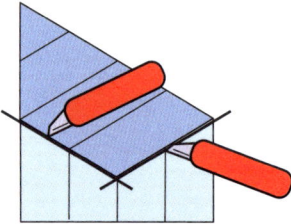

4

Make two cuts as shown, trimming away the single layer of paper below the double layer.

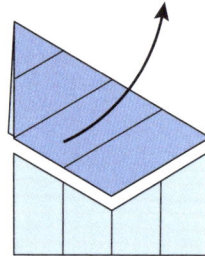

5

Open out the paper.

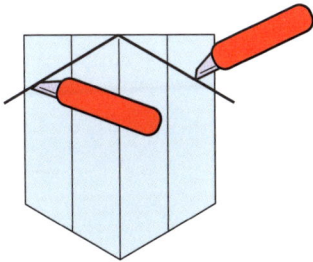

6

Open the triangles at the top
of the paper, then cut them off.

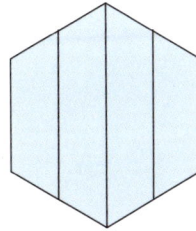

7

Complete. The hexagon is
pleasingly quick, easy and
direct to make, and there is
little waste from the paper
rectangle.

2.11 Heptagon from a Square

Heptagons are rarely seen, perhaps because they don't tile a surface and they are not used to create the more basic three-dimensional solids. It is this rarity that makes them enigmatic and has earned this little-known seven-sided figure a place in this book.

The example shown here is by the origami master Shuzo Fujimoto (Japan). Like the method to construct the pentagon, it is not wholly accurate, but the margins of error are so small that it can be regarded as a regular figure. The sequence is long, so fold carefully.

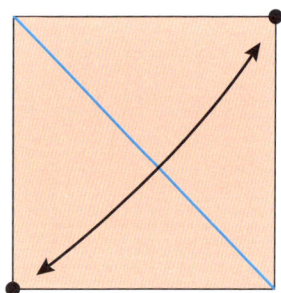

1

Fold corner to corner. Unfold.

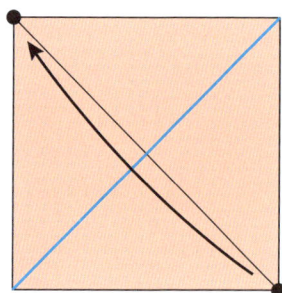

2

Fold the other two corners together, but this time, do not open the fold.

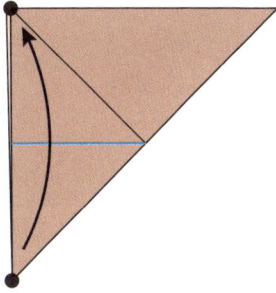

3

Fold a 45° corner to the 90° corner.

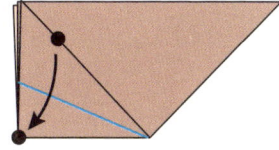

4

Fold down the sloping edge of the triangle to lie exactly along the bottom edge.

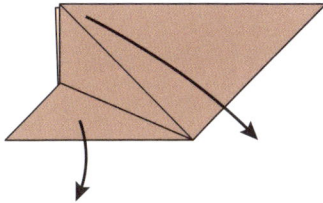

5

Open the paper completely.

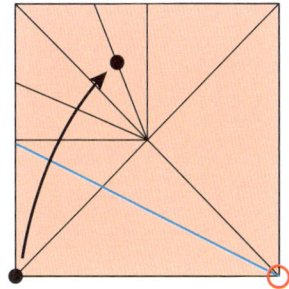

6

Fold up the bottom corner to lie along the open crease, as shown. Note that the fold must run *exactly* into the other corner at the bottom of the paper square. Take your time.

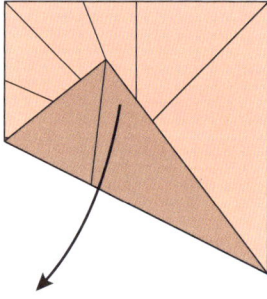

7

Unfold the triangle.

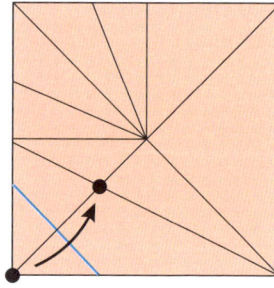

8

Fold dot to dot, as shown. This is the critical fold that enables the paper to be divided into seven equal angles.

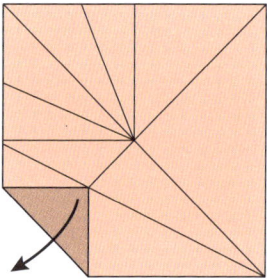

9

Unfold the triangle.

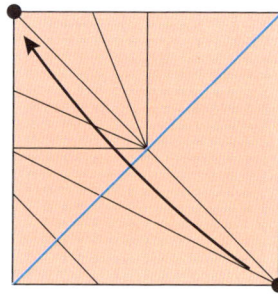

10

Refold step 2.

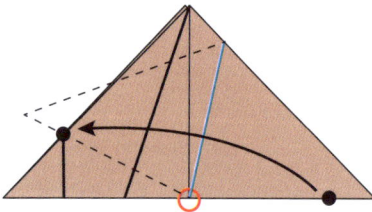

11

Fold dot to dot. Note that the fold originates at the centre point of the paper square. The dot on the right shows that the folded edge of the paper touches exactly the top of the crease you made in step 8. This is a very unorthodox fold that must be made accurately.

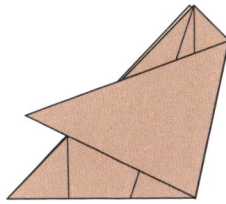

12

This is the result. Study it well before continuing. Turn over.

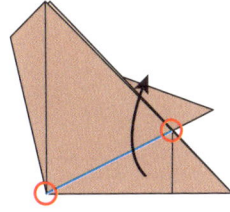

13

Fold as shown, so that the fold lies exactly on top of the edge underneath.

14

Fold the left-hand edge to butt against the folded edge. Note how all these folds are terminating at the bottom corner, which is becoming increasingly acute.

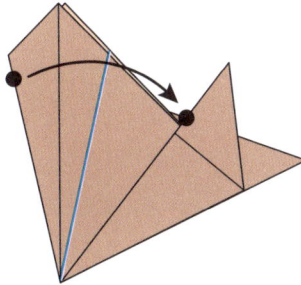

15

Check your result, then turn over.

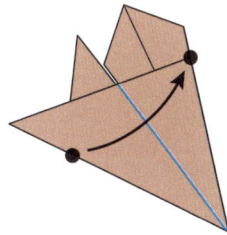

16

Fold in half.

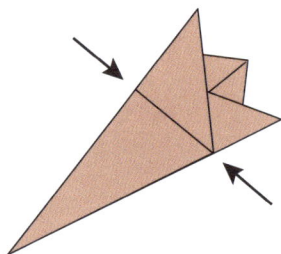

17

The arrows show the position of an existing crease.

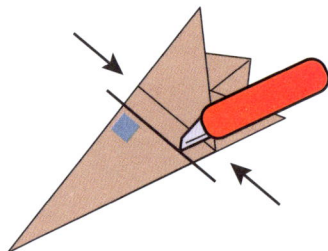

18

Using that crease as a guide, make a cut that is parallel to it, a few millimetres below it.

IMPORTANT The cut meets the edge at the left at 90°. This cut must be made precisely. You may wish to regard your first attempt at completing the heptagon as a dry run so that you can understand exactly what to do. The second attempt, folded with foresight, will then be perfect.

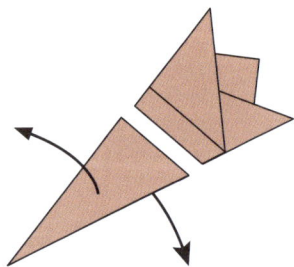

19

Discard the top portion and unfold the bottom portion.

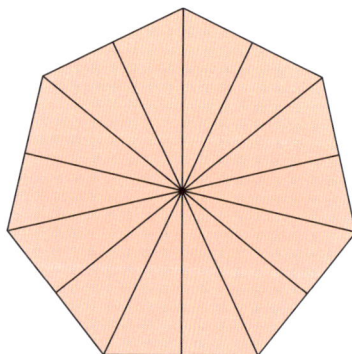

20

Complete. This is the heptagon. The folding sequence is lengthy, but if you follow it carefully, the result will be very accurate.

2.12   Octagon from a Square

Just as a hexagon uses the geometry of folding an equilateral triangle, so an octagon uses the geometry of folding a square. Since it is not possible to grow an octagon beyond a square and into a rectangle of paper, all the methods to make an octagon use square paper, never rectangular paper.

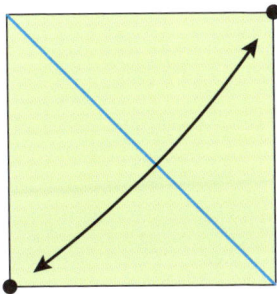

1

Fold one corner to the opposite corner. Unfold.

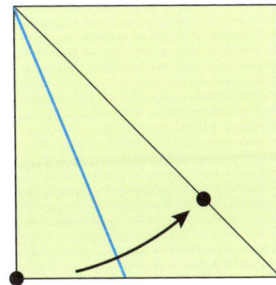

2

Fold a corner to lie on the open crease, such that the fold you are making runs exactly into a corner, as shown.

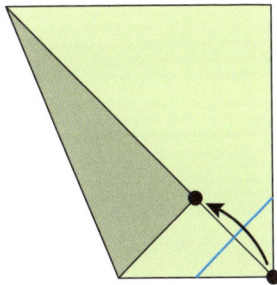

3

Fold dot to dot so that the corner of the square touches the corner of the triangle.

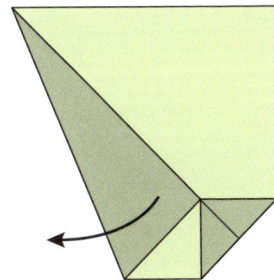

4

Unfold the larger triangle.

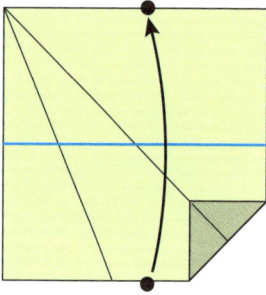

5

Fold the bottom edge of the
square up to the top edge.

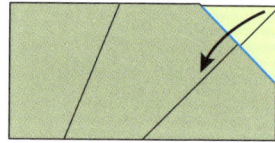

6

Wrap the single-layer corner
tightly over the folded edge.

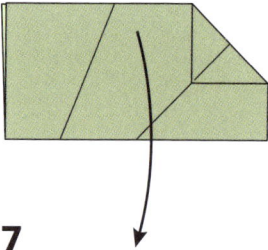

7

Unfold the crease you made
in step 5.

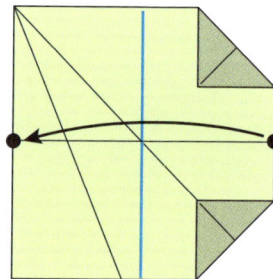

8

Fold the right-hand edge
across to the left-hand edge.

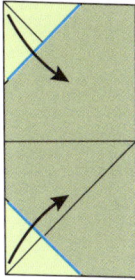

9

Wrap the two single-layer corners over the folded edges.

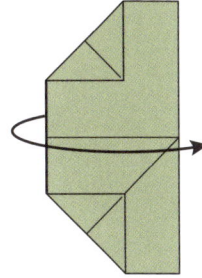

10

Unfold step 8.

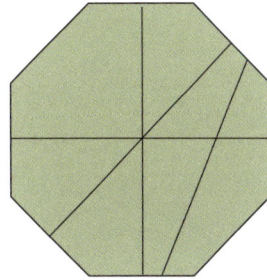

11

This is the final shape. Note how all the corners have been folded in by the same amount. Turn over.

12

Complete. There are many ways to determine the position of the creases across the corners of the square, and some are more direct than others. You may notice a similarity between steps 3–4 here and steps 3–4 in 4.2, which shows how to make an A4 (1:√2) rectangle from a square. The connection between 1:√2 rectangles and octagons may seem tenuous, but, as we shall see in Chapter 4, the connection is very strong.

③

Divisions

3
Divisions

Cataloguing the different ways of dividing paper into equal parts could take a whole book in itself. Many of these methods are elegant and use the technique of folding to great advantage, sometimes criss-crossing the sheet with many folds to create an exact division.

However, that is perhaps also the problem: these methods leave a sheet full of unsightly and usually unwanted creases. What is needed is a minimally invasive method, one that leaves few folds on the sheet.

This is achieved by using Shuzo Fujimoto's remarkable method of finding divisions using only one edge of a sheet of paper, not the full plane of a sheet. A series of pinches along a chosen edge, made by folding the corners to and fro in a specific sequence for each division (thirds, fifths, sevenths ...), leaves the paper looking clean. This chapter presents that beautiful method.

3.1 Thirds from an Edge: Slow Method

scan for video

This is the simplest of the pinch sequences. If you work through it diligently so that you can find thirds fluently, you will find the later examples that divide the paper into five, seven and nine easy to understand.

The method shown here shows the slow, theoretical way to create thirds along an edge. The next section shows how the method can be speeded up for practical use.

To make these examples, you don't need a sheet of paper, just an edge torn from a sheet. Once you understand the method, you can apply the technique to a full sheet.

1

Here is a sheet of paper with just one straight edge. It has two corners, marked LEFT and RIGHT. You might wish to write them on the paper.

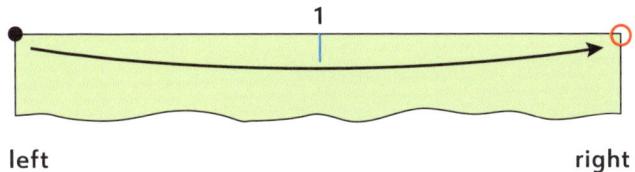

Pick up the LEFT corner and place it on the RIGHT corner, creating a short pinch at the midpoint of the edge. Open the paper. This is pinch no. 1.

left   right

2

Pick up the RIGHT corner and place it on pinch no. 1 to create pinch no. 2. Open the paper.

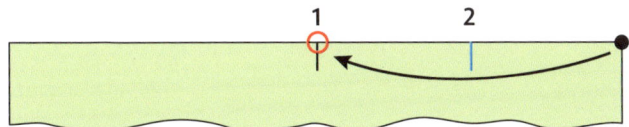

3

Pick up the LEFT corner and place it on pinch no. 2 to create pinch no. 3. Be sure to follow the sequence carefully.

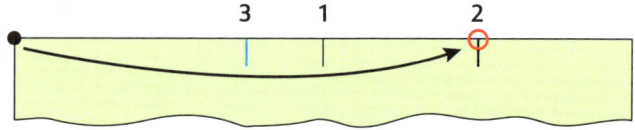

4

Pick up the RIGHT corner and place it on pinch no. 3 to create pinch no. 4.

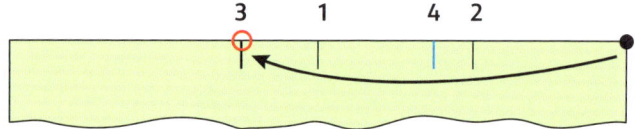

5

Pick up the LEFT corner and place it on pinch no. 4 to create pinch no. 5.

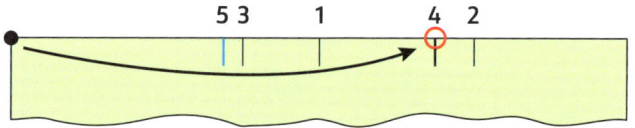

6

Pick up the RIGHT corner and place it on pinch no. 5 to create pinch no. 6.

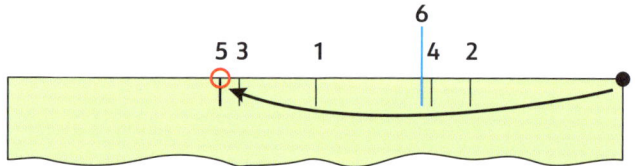

7

Pick up the LEFT corner and place it on pinch no. 6 to create pinch no. 7. Are you seeing the pattern of pinches? The corners are picked up in a strictly alternating sequence, left, right, left, right and so on.

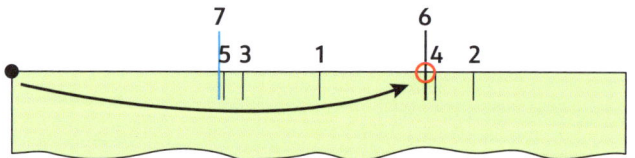

8

Pick up the RIGHT corner and place it on pinch no. 7 to create pinch no. 8.

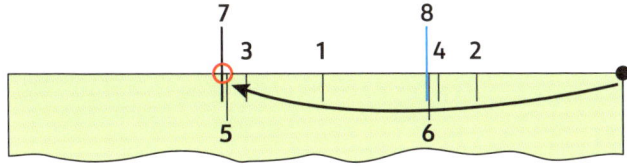

9

Pick up the LEFT corner and place it on pinch no. 8 to create pinch no. 9.

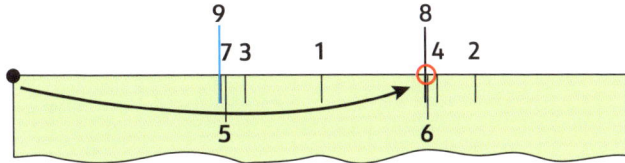

10

Pick up the RIGHT corner and place it on pinch no. 9 to create pinch no. 10. At about this point, after making 8–10 pinches, you may feel under your finger that you are making a pinch on top of a pinch that already exists. That crease line is exactly one-third of the way along the edge of the paper. You have found thirds.

We should now stop and recap what has happened. There are three important rules to understand:

Rule 1. The pattern of making pinches is to pick up the left corner, then the right corner, then the left, then the right ... and so on and so on, *always* alternating the corners, left, right, left, right ...

Rule 2. The position of a new pinch is *always* determined by bringing the corner of the paper *to touch the pinch just made*, then pinching the paper flat.

Rule 3. Rules 1 and 2 continue until the next pinch wants to be made on top of an existing pinch. This is the one-third mark on the paper's edge.

You will notice, as more and more pinches are made, that they begin to converge around the one-third and two-thirds divisions, until sufficient pinches are made (typically nine, ten or eleven pinches) and an exact third is found. Essentially, thirds are found by continually folding the paper in half from the left-hand side, then the right-hand side, then the left and so on.

With practice, thirds can be found in less than twenty seconds and with minimal creasing to the paper.

We can continue to the end.

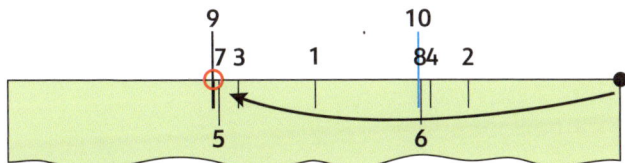

11

At around the tenth pinch, the next pinch will want to be made on top of an existing pinch. This is the one-third mark, so a fold can be made down the paper.

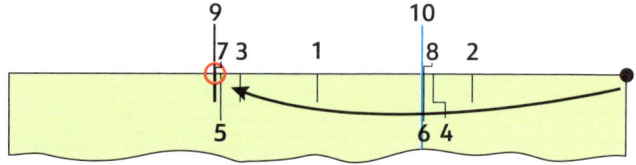

12

Fold the other corner around the back of the paper to create a zigzag like a letter Z.

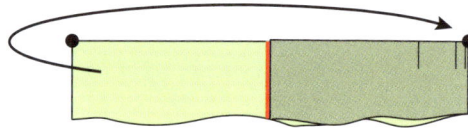

13

The paper should be folded into exact thirds. Unfold.

14

When the technique is copied on to a square or any rectangle, the sheet will be divided into perfect thirds. Dividing also into thirds down the side of the paper will create a grid.

The same technique can be used to divide any edge on any polygon, such as the edge of a triangle or pentagon.

In summary, we can describe the finding of thirds to be folding first the LEFT corner to create a pinch, then folding the RIGHT corner to that pinch, then repeating the left/right sequence until a pinch is made on top of an existing pinch. The pattern can be annotated as follows:

(L,R) ... (L,R) ...

The methods of creating other divisions, such as fifths and sevenths, will use different sequences of left and right pinches.

3.2 Thirds from an Edge: Quick Method

Although it works very well, the method described in 3.1 is very slow, requiring about ten pinches to locate a division of one-third. The method presented here is exactly the same, but instead of beginning with the first pinch at the midpoint of the edge, the first pinch is *estimated* to be at one-third. With practice, it dramatically reduces the time needed to create a division of a third to just a few seconds.

1

As before, begin by picking up the LEFT corner. However, rather than folding it to touch the right corner as before, instead drop it at a distance you estimate to be one-third of the length of the edge. The estimate need not be very accurate, but the more accurate it is, the fewer pinches you will need to find an exact third. Make a pinch, here marked 1.

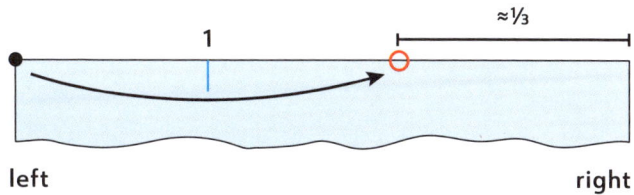

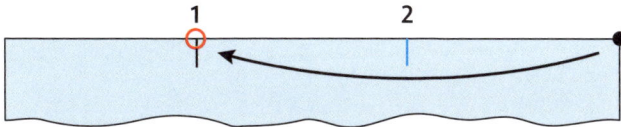

$\approx\frac{1}{3}$

1

left

right

2

Again as before, pick up the RIGHT corner and fold it to pinch no. 1 to create pinch no. 2.

1

2

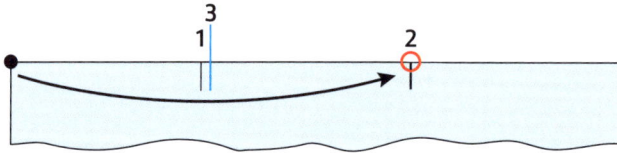

3

Again as before, pick up the LEFT corner and fold to pinch no. 2 to create pinch no. 3. Depending on the accuracy of the first pinch, pinch no. 3 may already be on top of no. 1, but you will probably need to make a few more pinches before you find the exact third.

For sure, this convergent technique is unusual and will take a little learning, but once learned, it becomes the quickest and least invasive way of dividing paper. The following pages adapt the simple left/right pinching pattern for creating thirds to create more complex divisions.

3.3 Fifths

The method used to create thirds is adapted here to create fifths. If you did not make the thirds example, please do so before you attempt to create fifths. The (L,L,R,R) ... (L,L,R,R) ... pattern of pinches is more complex than the (L,R) ... (L,R) ... pattern used to create thirds, but the convergence to find the perfect fifth is quick.

1

When using the convergent method to divide paper, the crucial fold – whatever the number of divisions needed – is a power of 2, namely, 2, 4, 8, 16 ... To find fifths, the important division is four-fifths (or one-fifth – it's the same division, but at the other end of the paper), because 4 is 2^2.

If we can find this, we can subdivide four-fifths to two-fifths by folding it in half, then halving it again to create one-fifth. By halving as many times as possible down to a single division, it becomes simple to infill the missing divisions and create a full set.

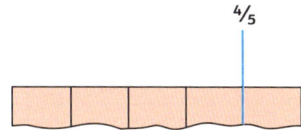

2

Pick up the LEFT corner and fold it to a point you estimate to be one-fifth of the edge away from the RIGHT corner. The more accurate the estimate, the fewer pinches you will need to make. Make pinch no. 1.

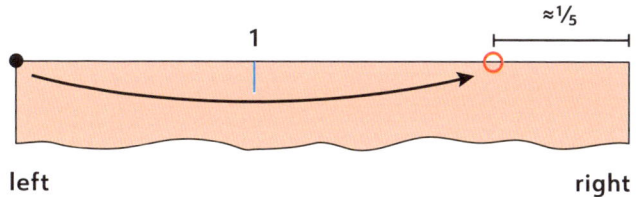

3

Again, pick up the LEFT corner and put it on pinch no. 1 to make pinch no. 2.

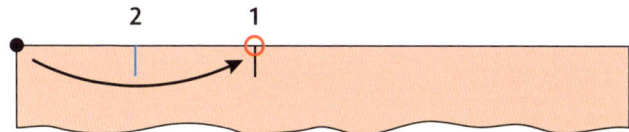

4

Pick up the RIGHT corner and put it on pinch no. 2 to make pinch no. 3.

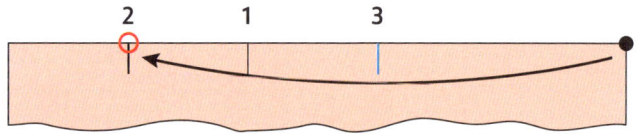

5

Again, pick up the RIGHT corner and put it on pinch no. 3. This is pinch no. 4.

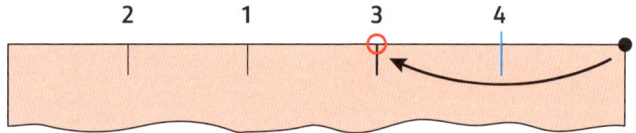

6

Repeat the (L,L,R,R) four-pinch cycle. Pick up the LEFT corner and fold to pinch no. 4. Make pinch no. 5. Continue the cycle as described above.

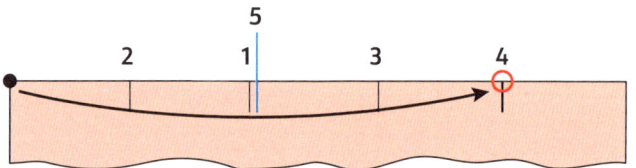

7

You will quickly feel that you are making a pinch on top of an existing pinch. If need be, make one more pinch so that a long fold can be made across the paper at the four-fifths or one-fifth mark. Fold down, across the sheet. This is your first creased division.

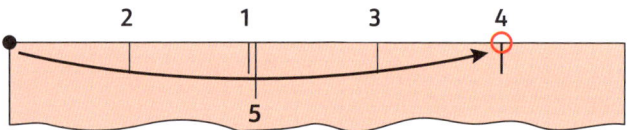

8

Here is the first accurate division at the four-fifths mark.

9

Find the two-fifths division by picking up the LEFT corner and placing it on the four-fifths crease, as shown. Make a fold.

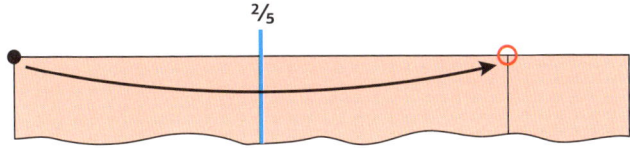

10

Find the one-fifth division by picking up the LEFT corner and placing it on the two-fifths crease, as shown. Make a fold.

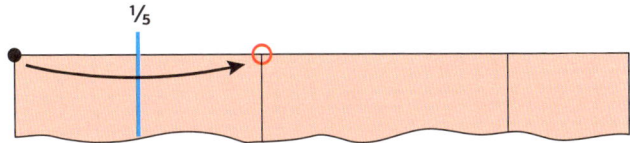

11

Find the three-fifths division by picking up the RIGHT corner and placing it on the one-fifth crease, as shown. Make a fold.

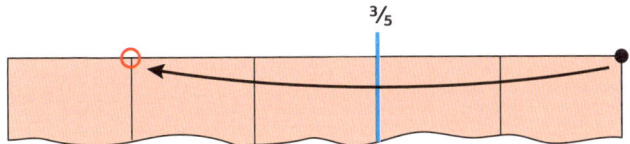

12

We have now found all the divisions to divide the edge accurately into fifths.

Note that whereas the thirds pinch cycle was a two-pinch (L,R) ... cycle, the fifths pinch cycle is a longer four-pinch (L,L,R,R) ... (L,L,R,R) ... cycle. Each division has its own unique L,R pinch pattern.

3.4 Sevenths

Pleasingly, to divide an edge into sevenths, we do not need to make all six pinches along the edge. We need make only three before the cycle repeats. The repeat cycle is (L,R,R) ... (L,R,R) ...

1

As always, the key to creating the cycle of pinches is first to find the division that is a power of two, but less than the number of divisions. Wishing to find sevenths, the important division to find is therefore four-sevenths (since 4 is 2^2). Four-sevenths will halve to create two-sevenths, which will halve again to create one-seventh, after which the other divisions can be filled in.

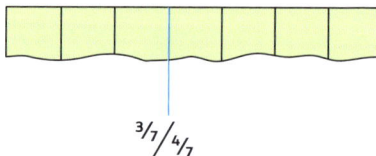

$^3/_7$ $/$ $^4/_7$

2

Pick up the LEFT corner and fold it to a point you estimate to be one-seventh of the edge away from the RIGHT corner. The more accurate the estimate, the fewer pinches you will need to make. Make pinch no. 1.

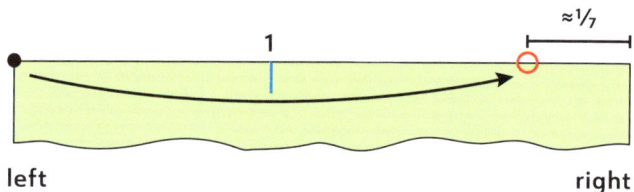

$\approx 1/_7$

1

left

right

3

Pick up the RIGHT corner and put it on pinch no. 1. Make pinch no. 2.

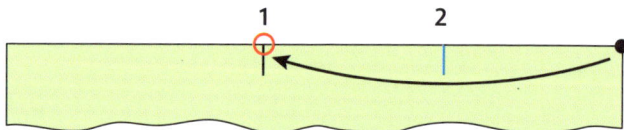

1

2

4

Again, pick up the RIGHT corner and put it on pinch no. 2. Make pinch no. 3.

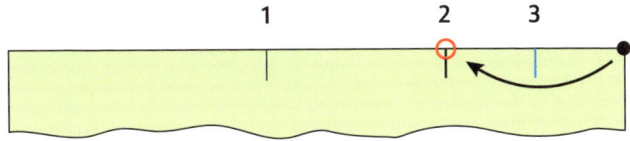

5

Repeat the (L,R,R) three-pinch cycle. Pick up the LEFT corner and fold to pinch no. 3. Make pinch no. 4. Continue the cycle as described in steps 2–4.

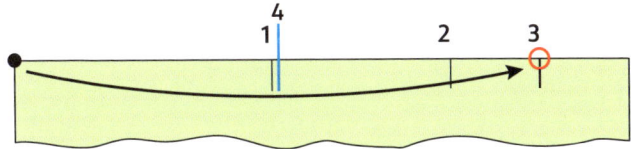

6

You will quickly feel that you are making a pinch on top of an existing pinch. If need be, make one more pinch so that a long fold can be made across the paper at the three-sevenths or four-sevenths mark. Fold down, across the sheet. This is your first creased division.

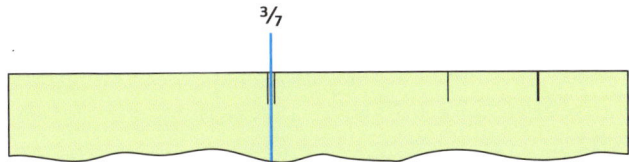

7

Find the five-sevenths division by picking up the RIGHT corner and placing it on the three-sevenths crease, as shown. Make a fold.

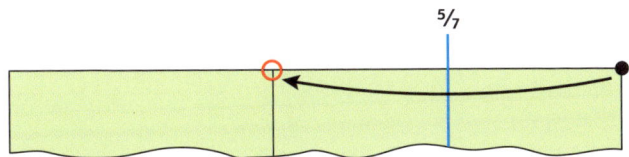

8

Find the six-sevenths division by picking up the RIGHT corner and placing it on the five-sevenths crease, as shown. Make a fold.

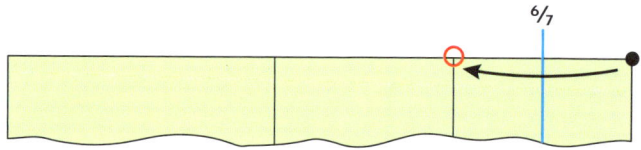

$^6/_7$

9

Find the four-sevenths division by picking up the five-sevenths crease and placing it on the three-sevenths crease, as shown. Make a fold. This is a little awkward, so take your time.

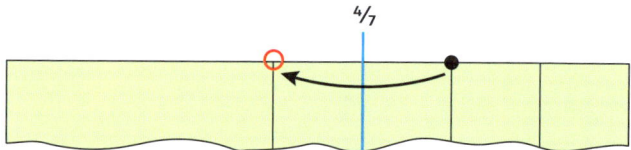

$^4/_7$

10

Find the two-sevenths division by picking up the LEFT corner and placing it on the four-sevenths crease, as shown. Make a fold.

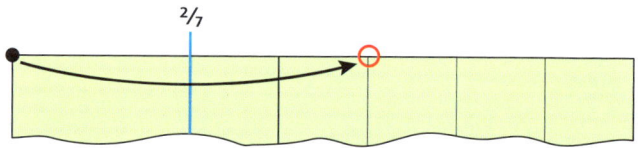

$^2/_7$

11

Find the one-seventh division by picking up the LEFT corner and placing it on the two-sevenths crease, as shown. Make a fold.

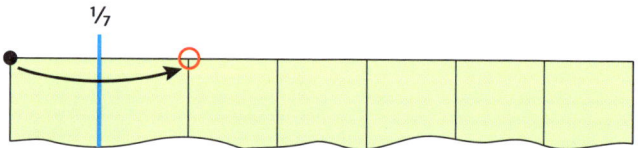

$^1/_7$

12

You have now found the complete set of divisions. This (L,R,R) ... (L,R,R) ... cycle of pinches is exceptionally economical.

3.5 Ninths

scan for video

As when dividing into sevenths, we do not need to create a full set of pinches in the ninths cycle. Instead of eight pinches, we need to create just six. The pattern is (L,L,L,R,R,R) ... (L,L,L,R,R,R) ... As before, we need to think which division is a power of two, but less than nine. The answer is 8 (2^3), so the important division to find is eight-ninths, which will divide in half to give four-ninths, then two-ninths, then one-ninth, after which the missing divisions can be filled in to create a full set of ninths.

1

The division of eight-ninths is the one to find.

2

Begin by picking up the LEFT corner and placing it on the top edge an estimated distance of one-ninth from the right corner. This is a difficult division to estimate accurately, but, since accuracy not important, give it your best guess. Make a pinch.

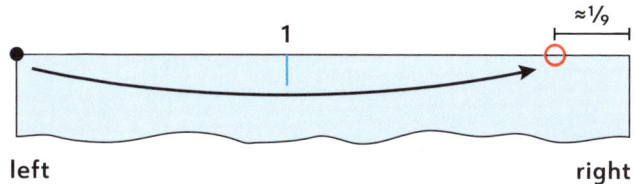

3

Again, pick up the LEFT corner and fold it to the pinch you have just made. Make a pinch.

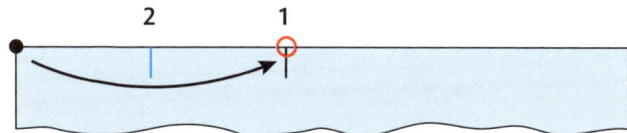

4

Again, pick up the LEFT corner and fold it to the pinch you have just made. Make a pinch.

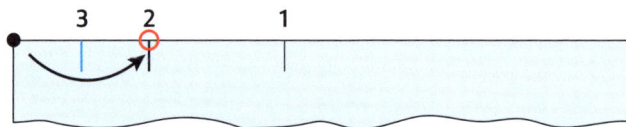

5

Pick up the RIGHT corner and fold it to the pinch you have just made. Make a pinch.

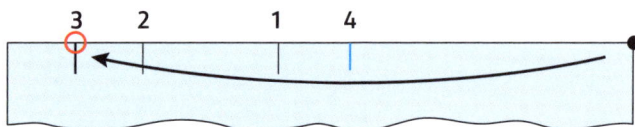

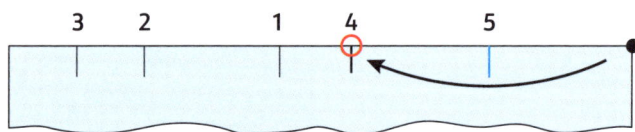

6

Again, pick up the RIGHT corner and fold it to the pinch you have just made. Make a pinch.

7

Again, pick up the RIGHT corner and fold it to the pinch you have just made. Make a pinch. This completes the first cycle of six pinches, (L,L,L,R,R,R).

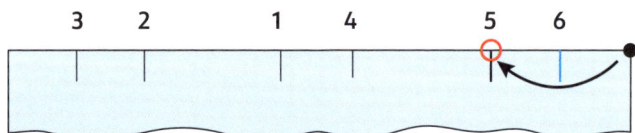

8

Begin to repeat the cycle, starting at step 2. Continue until you can feel an eight-ninths or one-ninth pinch under your finger that has already been made. This will be a perfect ninth.

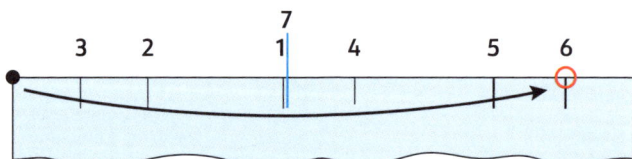

9

When you have found the perfect ninth, crease all the way down the paper.

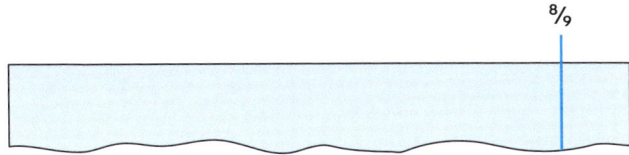

10

We can now begin to create the other divisions. Fold the LEFT corner to the eight-ninths fold and make a fold. This is four-ninths.

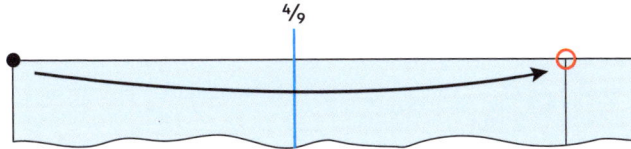

11

Fold the LEFT corner to the four-ninths fold and make a fold. This is two-ninths.

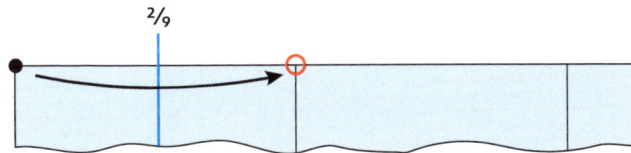

12

Fold the LEFT corner to the two-ninths fold and make a fold. This is one-ninth.

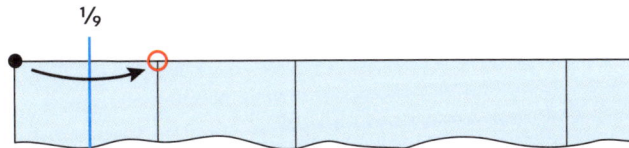

13

Fold the RIGHT corner to the one-ninth fold and make a fold. This is five-ninths.

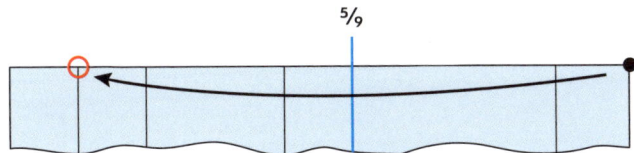

14

Fold the RIGHT corner to the five-ninths fold and make a fold. This is seven-ninths.

$^7/_9$

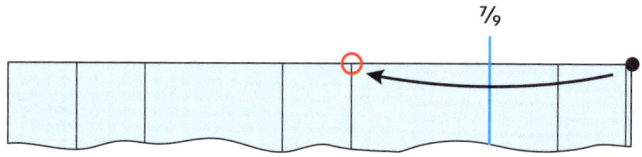

15

Fold the seven-ninths fold to the five-ninths fold and make a fold. This is six-ninths (or two-thirds).

$^6/_9$

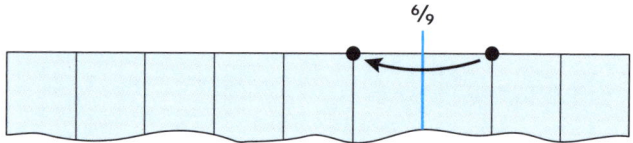

16

Fold the LEFT corner to the six-ninths fold and make a fold. This is three-ninths (or one-third).

$^3/_9$

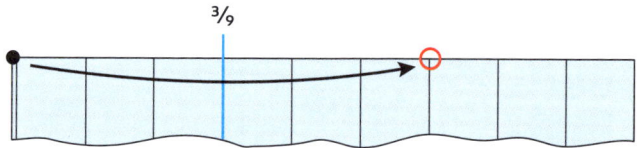

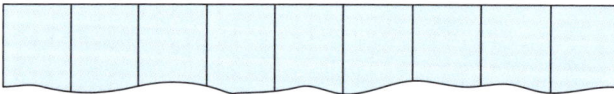

There are unique L,R patterns for any division, although the higher the number, the more esoteric the cycles become. Those cycles given on the previous pages should be sufficient for almost any purpose.

Dividing by folding is quicker and more accurate than dividing an edge using a ruler and pencil. Imagine having to divide the 297mm longer side of an A4 sheet of paper by 5 or 7 – the maths would be excruciating and liable to error. However, by folding it becomes easy and reliable.

3.6   Dividing an Angle

Remarkably, the pinch method of dividing an edge presented in this chapter can be used to divide any angle into any number of equal smaller angles. Exactly the same L,R patterns of pinches are made as have been described earlier in this chapter.

1

In this example, the angle at the bottom of the paper will be divided into three equal parts. Begin by picking up the LEFT corner and laying it on a point you estimate to be one-third of the distance from the RIGHT corner. Make a pinch. This is pinch no. 1.

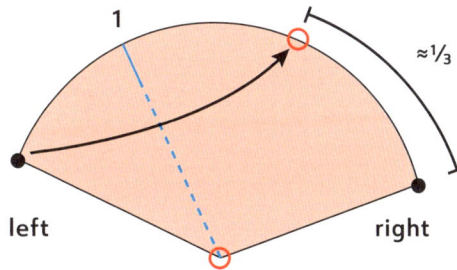

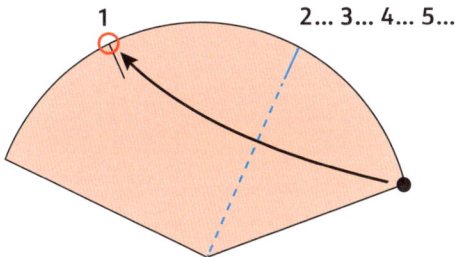

2

Pick up the RIGHT corner and lay it on top of pinch no. 1, making pinch no. 2. Repeat the L,R pattern until you make a pinch on top of a pinch. This method of trisecting the angle exactly copies the method of trisecting an edge presented in 3.1.

3

The angle has been trisected. Beautiful!

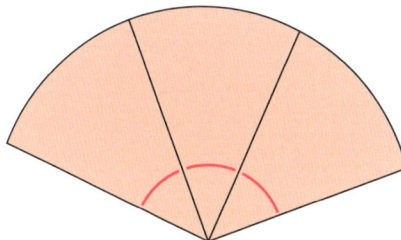

3.7 Halving a Given Division

Finding a division such as thirds or fifths is not necessarily the end of dividing the paper. Indeed, it may be only the beginning of a lengthy process. You can create a greater number of divisions simply by folding a division in half and in half again as many times as necessary.

1

Begin with a basic division, such as thirds. When each division is folded in half, thirds become sixths. Folded in half again, sixths become twelfths. Folded in half again, twelfths become twenty-fourths. Given a sheet of paper of sufficient size, there is theoretically no limit to the number of times this process can be repeated.

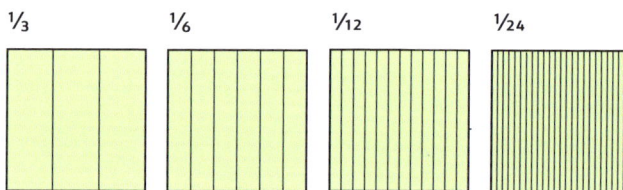

2

Divisions can be made as either columns down a sheet or rows across a sheet, or as both columns and rows to create a grid. A grid need not have the same number of divisions in its columns and rows. For example, a 3 × 3 grid could be 3 × 6, 12 × 3, 6 × 24 or 24 × 24 ..., as required.

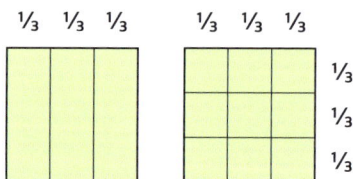

3

To create further divisions, all the existing folds on the sheet must be mountain folds. If they are not all mountains, either fold the valleys in the opposite direction to create mountains, or turn the sheet over so that the valley becomes a mountain. We can label the creases at this first level of division as no. 1 folds.

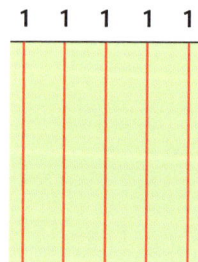

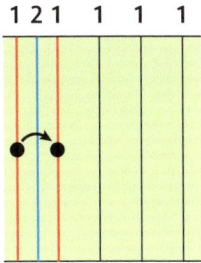

```
1 2 1   1   1   1
```

```
1 2 1 2 1   1   1
```

```
1 2 1 2 1 2 1 2 1
```

4

Gather up the first no. 1 mountain fold and lay it against the second no. 1 mountain fold. Flatten the paper to create a valley fold between them, which we can label as a no. 2 fold. Open the paper flat.

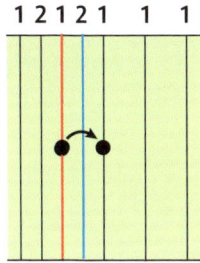

5

Pick up the second no. 1 mountain fold and lay it against the third no. 1 fold to create a second no. 2 fold. Open the paper out flat.

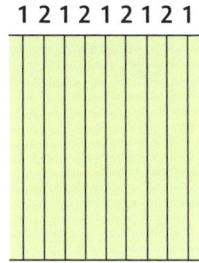

6

Repeat this process across the sheet. Remember always to fold a mountain fold to an adjacent mountain fold, and always open the paper flat after making each no. 2 fold.

7

All the no. 2 folds have been made. If the paper needs dividing again, repeat the process, creating folds between the no. 2 folds.

This technique will also work when the no. 1 folds are fifths, sevenths or ninths. It will also work folding the paper more conventionally in half, then into quarters, then eighths, then sixteenths ... and so on. By selecting the no. 1 division that best suits your needs, you can divide the paper into very many different divisions. For example:

HALF   divides into 4, 8, 16, 32, 64 ...

THIRDS   divides into 6, 12, 24, 48, 96 ...

FIFTHS   divides into 10, 20, 40, 80 ...

SEVENTHS divides into 14, 28, 56, 112 ...

NINTHS   divides into 18, 36, 72, 144 ...

4

Geometry from A4 Paper

4
Geometry from A4 Paper

A4 paper is arguably the most useful paper for creating folded geometry. Its proportion of 1:√2 enables the creation of many beautiful and practical constructions that are difficult or contrived using geometrically simpler square paper. This chapter focuses on some of the best and most poetic of the many possible constructions that make full use of its special shape, in both two and three dimensions. Many are very surprising and offer glimpses into the realm of geometry hidden in this most humble of everyday materials. If you read the chapter carefully and make as many of the examples as possible, you will never again look at A4 in quite the same way. This ubiquitous rectangle deserves our respect and repays our close attention, which is why this is the longest chapter in the book.

If you have only 8½ x 11in paper to hand, see 2.8 for a method of trimming it to 1:√2.

4.1   The What and Why of A4 Paper

A brief history

A4 paper has an aspect ratio between the short and long sides of 1:√2, or 1:1.414... This ratio was first adopted into law in France in 1798, in the aftermath of the French Revolution, and used alongside traditional ratios and sizes.

More than 100 years later, in 1911, the German advocacy agency Die Brücke (The Bridge) proposed that paper with the aspect ratio of 1:√2 be adopted as the world standard, to replace the plethora of confusing paper sizes in Germany, Europe and beyond. After long discussions, the ratio was adopted in 1921 by the Deutsches Institut für Normung (DIN; the German Institute for Standardization), becoming known as DIN 476.

DIN 476 was quickly taken up by many countries across the world. In 1975 it was adopted by the non-governmental International Organization for Standardization (ISO) as the world standard, and became known as ISO 216. Today, DIN A4 paper, or ISO A4 paper – or simply A4 paper, or just 'A4' – is common throughout the world, with the significant exception of the United States, where 8½ x 11in paper, known as American Letter Size paper, ALS or Letter paper, dominates. In Canada, Mexico and several South American countries, the use of either A4 or ALS paper depends on local preference.

In addition to the widely used ISO-A series of paper sizes, including A4, there are other series, all with the same 1:√2 aspect ratio. These include the B series and the C series. The B series is based on the geometric mean of the A series paper sizes. B0 (B-zero) paper has a shorter side of exactly 1m. The rarely used C series is the geometric mean between the A and B series paper sizes.

The proportion, size and weight of ISO papers

Proportion

As we have seen, all ISO papers of any size have an aspect ratio of 1:√2. Significantly, this is the ratio between the side of a square and its diagonal. This proportion is unique because it is the only one for which, when a sheet is folded or cut in half, the new halves will each have the same aspect ratio of 1:√2. When divided in half, a rectangle with a different aspect ratio will result in smaller sheets with a different aspect ratio from the undivided sheet.

Thus, whatever the size of the sheet, the aspect ratio of ISO sheets will remain the same. This is a commendably logical system that ensures a consistent proportion, whatever the size of the paper sheet.

1

If the short side of A4 paper has a unit length of 1, the long side will have a unit length of √2, or 1.414... This is the same ratio as the side of a square to its diagonal. Thus, a fold made at 45° to a corner of the A4 will have the same length as the long side.

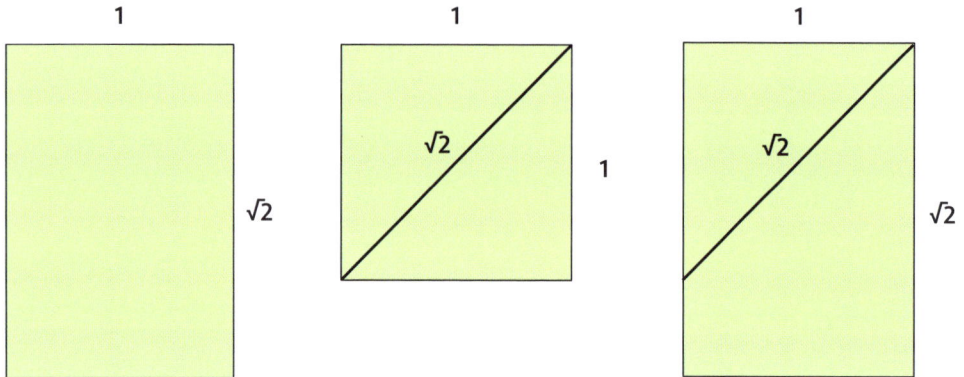

At about the time the French adopted the 1:√2 series of paper sizes, they also created the metric system of measurement. A metre was calculated to be one ten-millionth of the distance from the Equator to the North Pole.

Thus, the millimetre paper sizes of the DIN series, including A4, relate not only to the length of a metre, but also to the size of the Earth. That's quite a thought.

Geometry from A4 Paper

2

When cut in half, a sheet of paper proportioned 1:√2 will divide into two halves, each with the same 1:√2 proportion. This proportion will be retained each time a sheet is cut in half. This retention of a proportion does not happen with rectangles of other ratios.

Dimensions and areas

Two A4 sheets can be joined side by side (long edge to long edge) to create an A3 sheet, two A3s can be joined to create an A2, two A2s can be joined to create an A1, and two A1s can be joined to create an A0. The aspect ratio of A0 is 1:√2 and its size is 1,189 x 841mm. This appears to be an ungainly and random measurement, but 1,189 x 841 = 999,949, or almost 1,000,000 (one million). One million square millimetres = one square metre. So, a rectangular sheet of A0 paper has an area of 1 sq m, or 1m². It's another example of the beautiful rationality of the DIN paper series.

From this, we can calculate the areas of the ISO series, as follows:

ISO	Area
A0	1m²
A1	½m²
A2	¼m²
A3	⅛m²
A4	¹⁄₁₆m²

... and so on

From this table, we can see that the area of A4 paper is one-sixteenth of a square metre. To put it another way, sixteen sheets of A4 paper will cover 1 sq m.

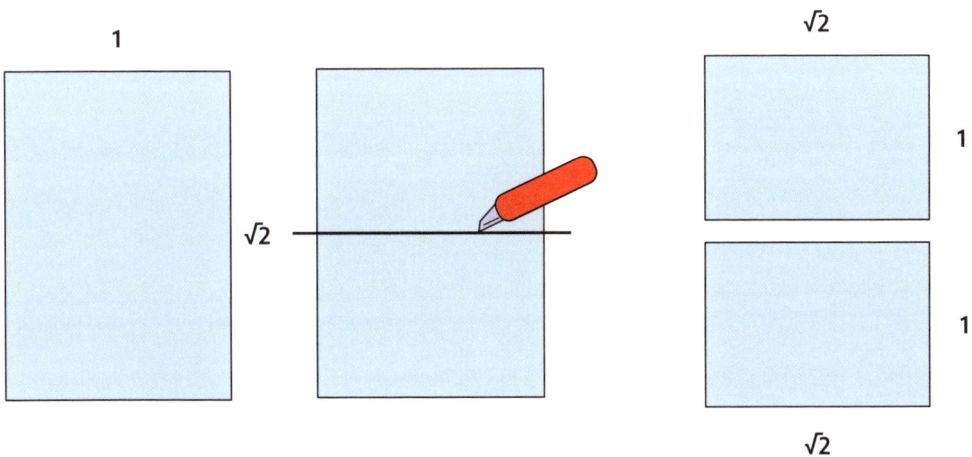

3

The family of A-sized papers is shown opposite. Each sheet is proportioned 1:√2 and each is either half the area or twice the area of its numerical neighbour. The smallest sheet labelled here is A7, but in theory, the A sizes could decrease in size until the paper was barely visible.

Weight

How much does a sheet of ISO paper weigh? If we know the grams per square metre (gsm or gm^2) of the paper, it is an easy calculation.

'Grams per square metre' means 1 sq m of that paper will weigh the specified number of grams. A packet of A4 copy paper will commonly have '80gsm' written on the wrapper and is often called 'eighty-gram paper'. This means that 1 sq m of copy paper will weigh 80g. Since we know that A4 is one-sixteenth of a square metre (see Dimensions and Areas on page 85), the weight of one sheet of A4 copy paper must be one-sixteenth of 80, or 5g. How cool is that? An A3 sheet of the same paper will weigh 10g, an A2 sheet will weigh 20g, and so on.

Conversely, if we know the size of the sheet but don't know the grammage, we can weigh the sheet and multiply it up to a square metre, using its size as a reference. The answer will often be an approximation, but accurate enough to be useful.

The joy of A4

The ISO paper series is the epitome of rational design. Using it, we can easily know the proportion, dimensions, area and weight of any sheet of paper. For a printing house, being able to calculate the weight of a booklet with – say – ten A4 sheets weighing 100gsm each and a card cover weighing 300gsm means being able to calculate the weight and therefore the cost of mailing individual copies, or 1,000 copies in a large carton, and, in turn, giving this information to the customer as part of the quotation for a job. Standardizing the size of printed papers also means that the sizes of envelopes and boxes can be standardized. Standardization simplifies production and reduces costs.

The other reason why ISO paper is a joy is its geometry. The aspect ratio of 1:√2 is arguably the most significant ratio in all folded geometry, relating as it does the side of a square to its diagonal. As you work through this chapter, you will encounter many beautiful examples of this magical ratio working to construct two-dimensional shapes and three-dimensional forms with simplicity and elegance.

A4 paper is so humble and ever-present that it is rarely given attention. But its proportion, dimensions and weight are interrelated in such a beautiful and logical way, and its geometry is so profound, that it should be regarded as one of the true design classics of modern times.

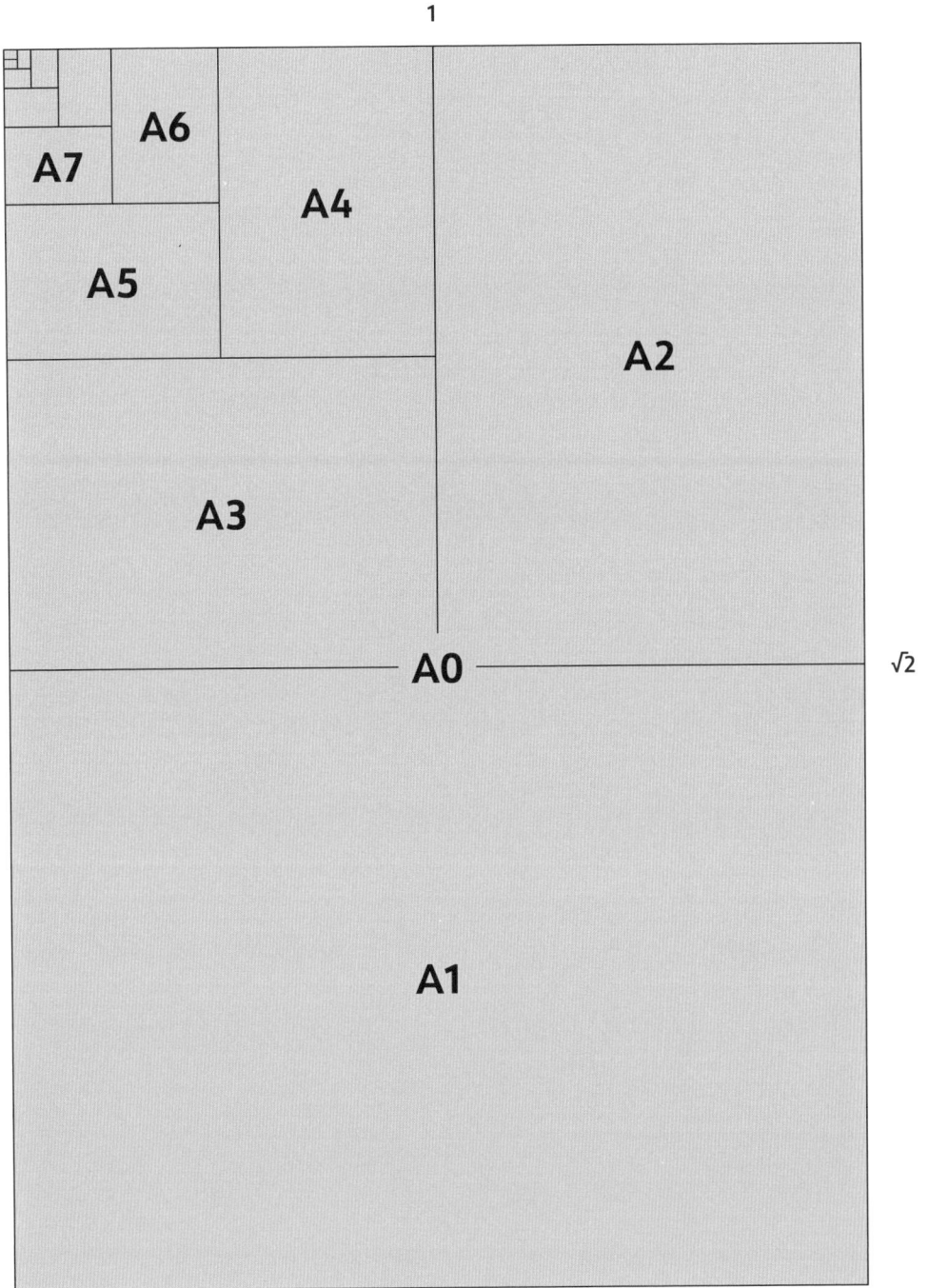

4.2   1:√2 Rectangle from a Square

If you have only square paper to hand – perhaps a packet of origami paper or a telephone notepad – here is an elegant way to convert it quickly to a 1:√2 rectangle.

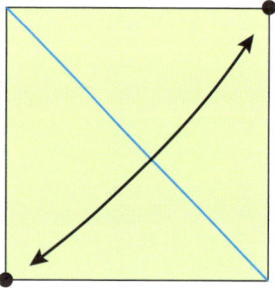

1

Fold a diagonal. Unfold.

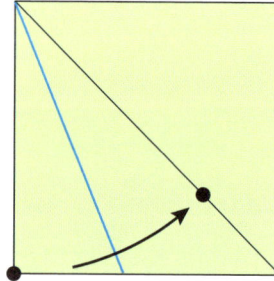

2

Bring an edge of the square to lie along the unfolded diagonal crease. The new fold will run exactly into a corner.

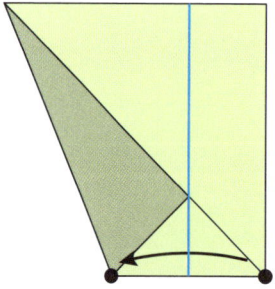

3

Note the short edge at the bottom of the paper square. Fold the corner of the square to the end of the crease just made in the previous step, as shown. The new fold will exactly touch the corner of the square on the diagonal crease. Nice!

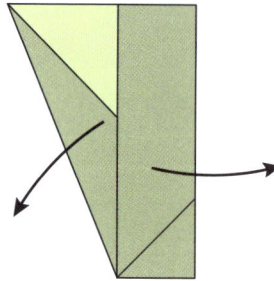

4

Unfold everything.

5

This is the complete crease pattern.

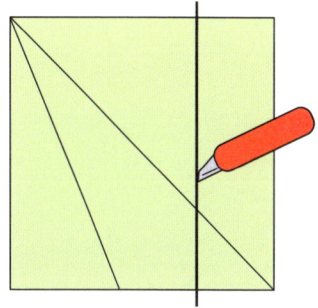

6

Cut carefully along the vertical crease.

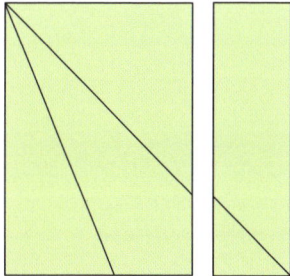

7

The square is now divided into two rectangles. The smaller rectangle can be discarded.

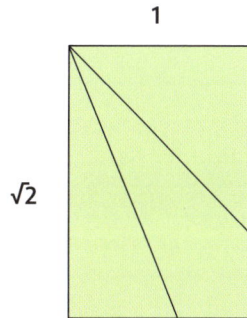

8

The larger rectangle has the proportion 1:√2. Can you see why it would have this proportion?

1

√2

4.3 Square from a 1:√2

This exquisite method of creating a square by Alessandro Beber (Italy) makes full use of the geometry of a 1:√2 rectangle. With just two small pinches, a square can be removed, leaving the square almost unmarked. The method is first shown using full-length folds, then repeated, using only small pinches.

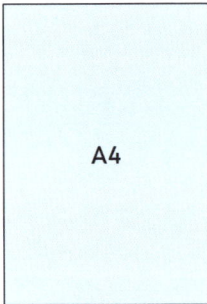

1

Use an A-series rectangle, such as A4.

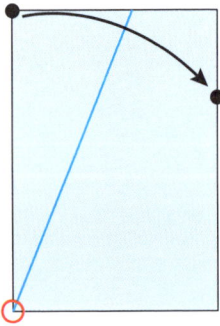

2

Look carefully. The top left corner is folded across to touch the right-hand long edge of the paper, so that the fold made runs exactly to the bottom left corner. Take your time to make this fold accurately.

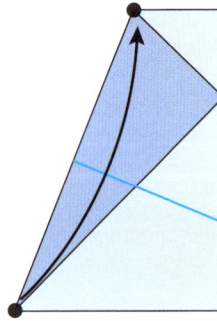

3

Make a new fold, so that one end of the fold just made touches the other end. The folded edge is being folded in half.

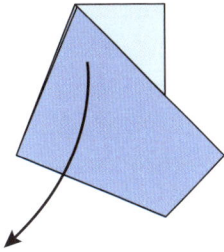

4

Unfold the paper completely.

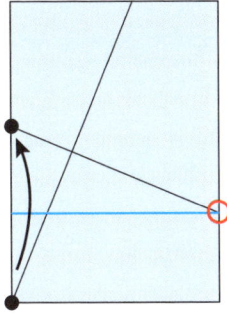

5

Fold dot to dot, as shown. Satisfyingly, the new fold will exactly touch the right-hand edge of the paper where the second fold also touches the edge.

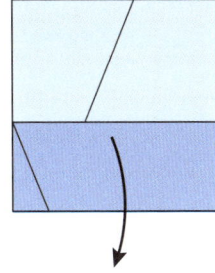

6

Unfold the previous step.

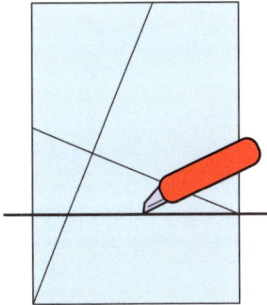

7

Cut along the horizontal crease, dividing the rectangle into two pieces.

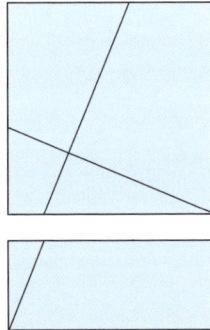

8

The larger piece is an exact square.

We will now repeat the folding sequence, but using only small, non-invasive pinches, instead of long folds.

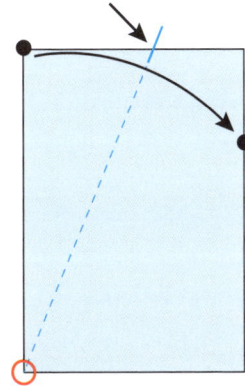

9

Fold the top corner, as shown. *Bend* the paper along the line of the fold, being careful not to make a crease line. When the corner is accurately placed on the right-hand edge, make a small pinch at the top edge, as shown.

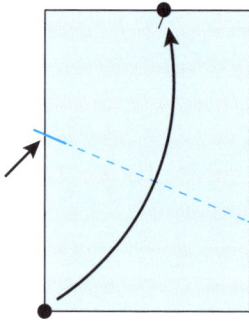

10

Bring the bottom corner up to the pinch, but do not make a fold. Instead, just make a small pinch at the left-hand edge, as shown.

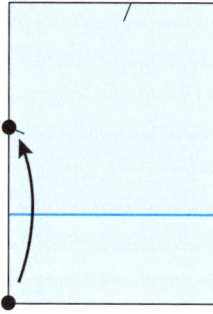

11

Bring the bottom corner up to the second pinch and fold across the paper.

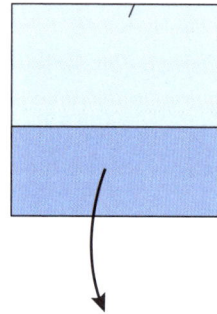

12

Unfold the paper.

13

Separate the square from the rectangle.

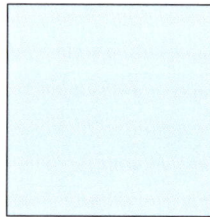

14

Here is the square. With a little practice, the two pinches can be made extremely small so that when the square is further folded, they become all but invisible.

The geometry of this construction is subtle. Even if you have little or no knowledge of geometry, logical reasoning alone will give you a proof of why what looks like a square truly is a square ... although the proof may take a few minutes to find.

The beauty, speed, minimalism and simplicity of the method make it a true classic. It is a great example of why folding paper is not just about the beauty of the result, but also about the beauty of the folding method. Folding paper should be a pleasure in itself, not simply a means to an end.

4.4 Isosceles Triangle from a 1:√2

An isosceles triangle has two equal angles and two equal sides. The method of making an example using 1:√2 paper is beautiful and exceptionally simple – almost absurdly so.

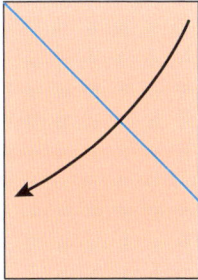

1

Take a 1:√2 rectangle and fold a triangle as shown.

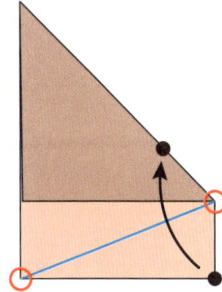

2

Look at the rectangle at the bottom. Fold up the short side at the right, so that it lies on the folded edge. When this is done, the new fold will run exactly to the corner at the bottom left. If the new fold does not want to go to the corner, move it slightly so that the alignment with the corner is accurate. Any inaccuracy is because of your folding, not because of the geometry.

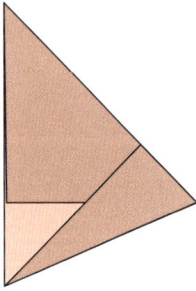

3

Turn the paper over.

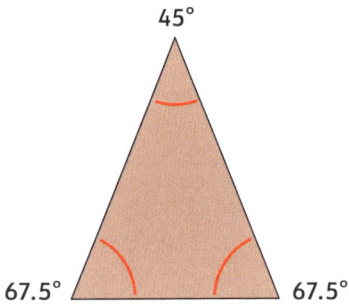

45°

67.5°   67.5°

4

Complete. What are the
angles of the isosceles
triangle? The angle at the top
was made by folding a 90°
corner in half, so it must be 45°.
Since there are always 180° in
a triangle, the other angles will
total 180 − 45, which is 135°.
These two angles are equal,
so they are both 67.5°.

This 45-67.5-67.5° isosceles triangle
has a strong relationship to an octagon,
an eight-sided polygon. If eight of
them are arranged in a circle with the
45° corners at the centre, they will form
a regular octagon.

This relationship between 1:√2 and an
octagon may seem unlikely, but it is
common, as the next few constructions
will demonstrate.

4.5 Isosceles Triangle from a 1:√2 Offcut

The narrow rectangle removed from a 1:√2 when a square is created looks random and of little use. However, it holds some compelling secrets, particularly in relation to constructing octagons.

The proportion of the offcut rectangle is 1:√2 + 1
(1 is the short side and √2 + 1 is the long side).

1

Here is a 1:√2 rectangle with the square separated from the offcut.

Fold one corner of the offcut to the opposite corner.

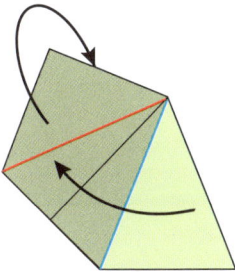

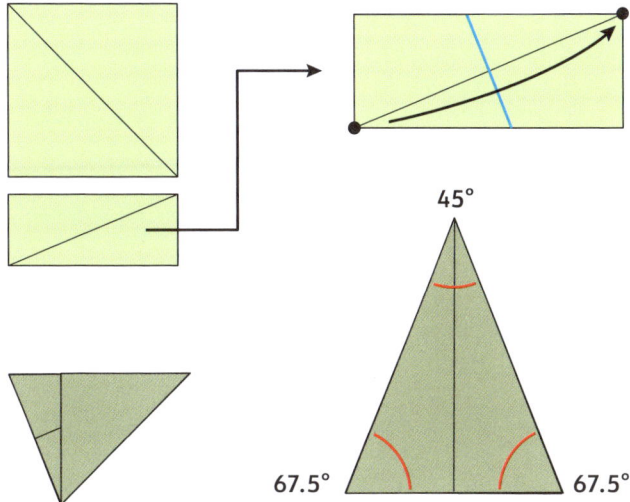

45°

67.5° 67.5°

2

Before folding further, notice that at this stage, the folded shape has three triangles; the outer triangles are one layer thick and have angles of 90-45-45°. The central triangle is two layers thick and has angles of 45-67.5-67.5°. This is entirely unexpected.

Wrap the single-layer triangles around the central double-layer triangle, using one valley and one mountain fold.

3

This is the result. Note that, as in 4.4, we have once again created a 45-67.5-67.5° triangle. Eight of them will make an octagon.

4

If you open the three folds a little, the front and back triangles can be relocated *inside* the layers of the 45-67.5-67.5° triangle such that they lock the construction shut and leave it looking clean on both sides. This very satisfying and clever lock, requiring no new folds, is a little like folding your arms across your chest. Can you discover how to make it?

4.6 Octagon from a Square and the Offcut

scan for video

This is perhaps the most remarkable construction in the book, revealing first, a profound relationship between the square cut from a 1:√2 and the remaining offcut, and second, the relationship they have to an octagon. Further, the construction is simple and direct. If you wish to show someone the geometry hidden in a common sheet of A4 paper, this is perhaps the example to show. It's stunning.

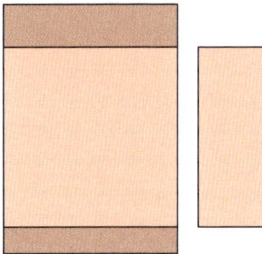

1

Use the method described in 2.6 to separate a square from the offcut without making any folds, using two identical rectangular sheets.

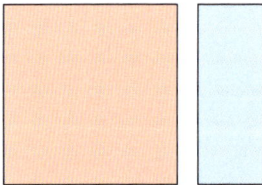

2

For clarity, although the two pieces have been cut from the same sheet of paper, they will be coloured differently here.

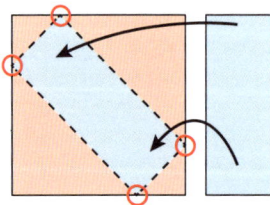

3

Lay the offcut across the square at 45°. Notice how the four corners of the offcut *exactly* touch the four sides of the square. Take time to arrange the offcut with precision. The relationship is entirely unexpected.

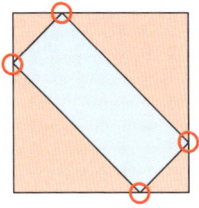

4

This is the relationship between the two pieces. Before you proceed, check that the alignment is exact.

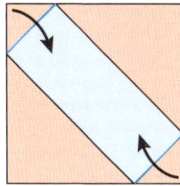

5

Fold over the corners of the square along the short edges of the offcut.

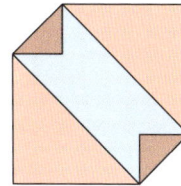

6

This is the result.

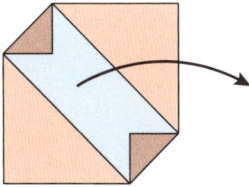

7

Take out the offcut.

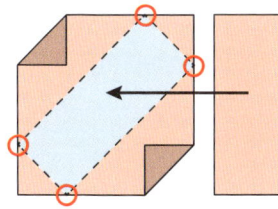

8

Lay the offcut on the square again, but this time along the second diagonal.

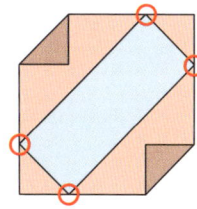

9

In the same way as before, align the corners of the offcut with the sides of the square.

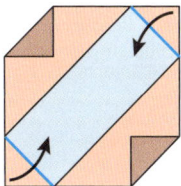

10

Again as before, fold the corners of the square over the short sides of the offcut.

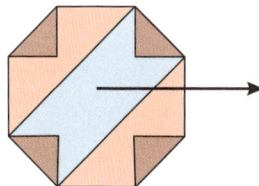

11

Remove the offcut.

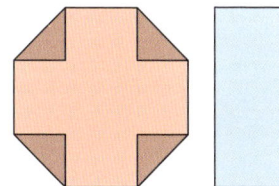

12

The construction is complete. The only folds you have made are those across the corners of the square to create the octagon.

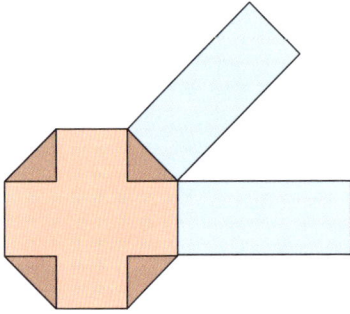

13

The polygon is a regular octagon. To prove it, lay the offcut against each of the sides. You will notice that all eight sides are exactly the same length as the short side of the offcut.

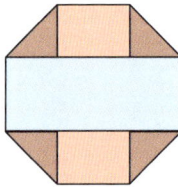

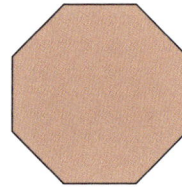

14

We know the offcut will span the diagonals between the triangles because this is how we folded the triangles across the corners of the square. Notice that it will also exactly span the raw edges of the square. This means that the shape of the offcut when a square is removed from a 1:$\sqrt{2}$ rectangle is the proportion of the rectangle that connects the opposite sides of an octagon.

Turn the paper over.

15

The octagon is one of the most well-known and best-understood polygons, but unfortunately, this beautiful method to construct one from a sheet of 1:$\sqrt{2}$ paper is little known. Once again, a prosaic result has been achieved using a poetic folding method.

Geometry from A4 Paper

4.7 No-Fold Octagon from the Offcut

In a book about folded geometry, it may seem a little perverse to include a construction that uses no folding at all, yet this delightful party piece fully merits its place, being related to the octagon on the previous pages.

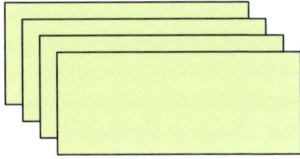

1

Take four sheets of 1:√2 paper, remove the squares and keep the offcuts.

2

Place one offcut horizontally.

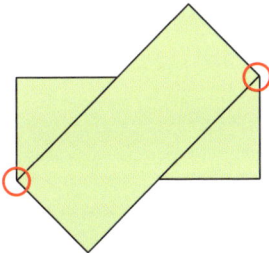

3

Place an offcut on top, as shown. Note that the corners are exactly touching. This is a simple alignment to make, yet it is somehow unintuitive, and it may take you a while to understand the relationship between the pieces.

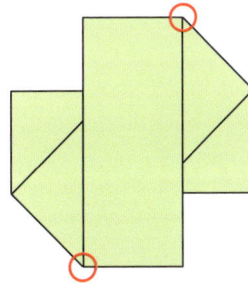

4

Similarly, align a third offcut, as shown.

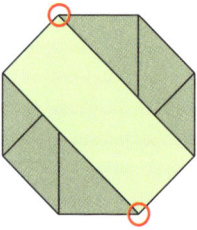

5

Finally, align the
fourth offcut.

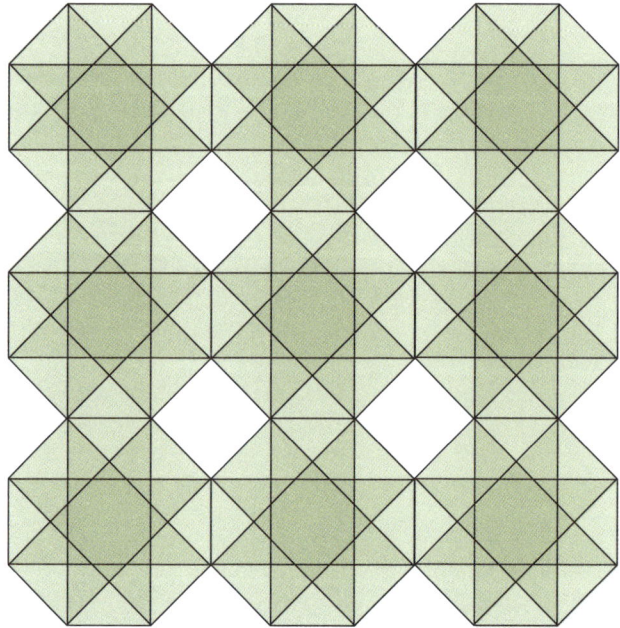

6

The four offcuts align to
make a regular octagon.
Held together with glue
and placed on a window
with strong backlighting,
a beautiful pattern is revealed,
created by the different
numbers of layers.

Several translucent octagons
placed together make a very
pleasing display.

4.8 Kite from a 1:√2

This is another elegant construction, requiring just two simple folds. As before, eight of these constructions will combine to create an octagon. Can you see how?

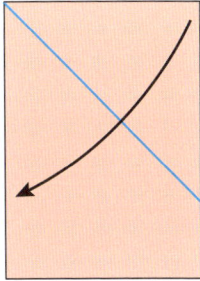

1

Take a 1:√2 rectangle and fold a triangle, as shown.

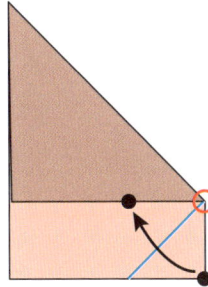

2

Look at the rectangle at the bottom. Create a small triangle at the end of the rectangle below the 45° corner on the large triangle, as shown.

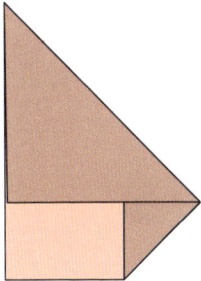

3

This is the result. Note that the small rectangle is proportioned 1:√2. The long side of the small triangle next to it will likewise have a length of √2. This is visual confirmation that the rectangle seen in 4.8.2 has the proportion 1:√2 + 1.

Turn the paper over.

4

Rotate the paper.

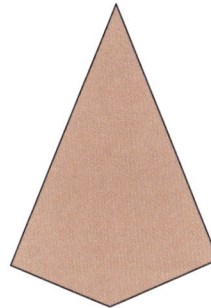

5

Surprisingly for many, the paper is symmetrical, having the shape of a kite (sometimes called a deltone or deltoid). In addition to an octagon, which other polygons or repeat patterns can be made from it?

4.9 Envelope

This is one of the few constructions in the book to have a clear use. The first fold is defined by a simple location, after which the other three folds follow in a straightforward manner. What is remarkable is the way these few folds create a complex web of halves, thirds and quarters across the sheet.

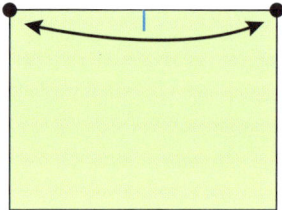

1

Pinch the midpoint of a long side, as shown.

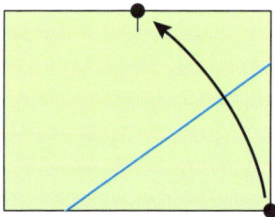

2

Pick up the bottom right corner and place it on the pinch. This fold is clearly defined, but it seemingly creates a random set of polygons and angles. We shall see at the end how this is not so.

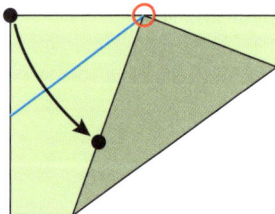

3

Fold down the top left corner, so that the top edge of the rectangle butts against the edge of the triangle. Notice how the two folds are parallel to each other and how the two triangles are similar (the angles are the same). Their relative sizes are in the proportion 1:$\sqrt{2}$.

4

Turn in the bottom left corner, as shown. Remarkably, it will exactly fill the space.

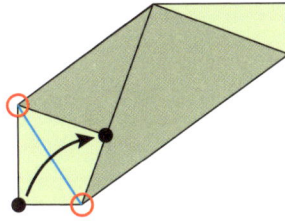

5

Finally, fold in the remaining corner of the rectangle so that it touches the other two corners. Notice how the fold exactly coincides with the midpoint pinch you made in step 1.

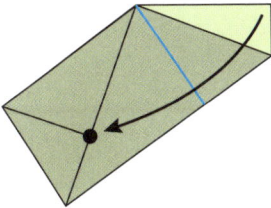

6

The folding is complete. Note that the envelope is a 1:$\sqrt{2}$. To use it as an envelope, unfold it and insert whatever it needs to contain.

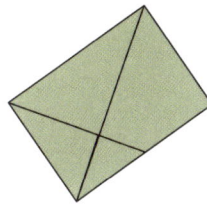

7

Use a sticker to secure the loose corners.

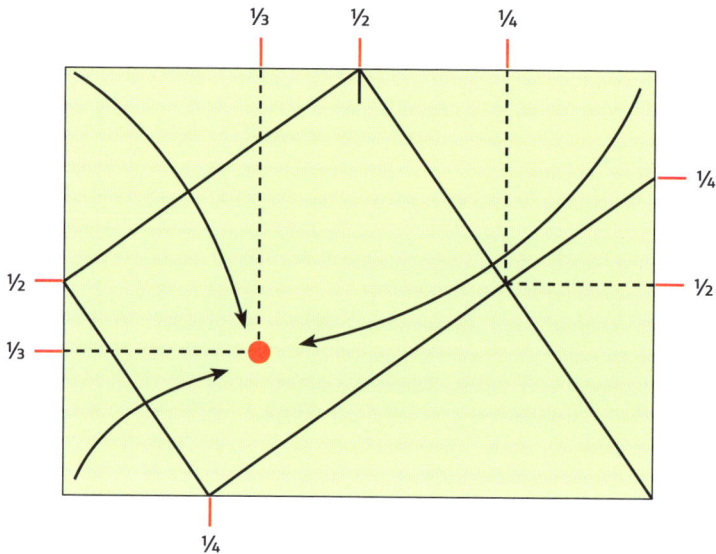

8

The crease pattern on the open sheet is well worth close study. From the simple fold you made in step 2, a sequence of simple folds has been made that divide the paper into a symphony of halves, thirds and quarters, both horizontally and vertically.

Of special interest is the red dot, the unmarked place where three corners of the rectangle come to rest, which – somewhat improbably – divides the rectangle into thirds. The dot also divides the folded envelope into thirds.

This complex construction can be done only with a 1:$\sqrt{2}$ rectangle. Any other rectangle folded in the same way will create little of geometric interest. It is remarkable that so much geometry can be derived from such a simple pattern of folds.

4.10   The Geometry of 1:√2 and Cubes

Looking at a flat 1:√2 rectangle of paper, its relationship to a three-dimensional cube is not immediately apparent. However, if we consider that another way to describe a 1:√2 rectangle is to say that the relationship between its two sides are the same as that of the side of a square to its diagonal, and since cubes are composed of square faces, the relationship becomes clearer. This short section describes this relationship between the two- and three-dimensional manifestations of 3-D in more detail, in preparation for the three-dimensional constructions that follow.

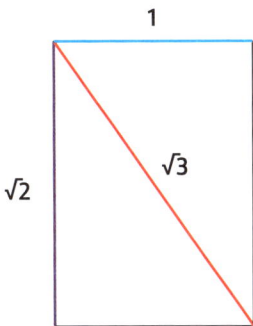

1

This is a 1:√2 rectangle. Note the diagonal, with a length of √3 (1.732...).

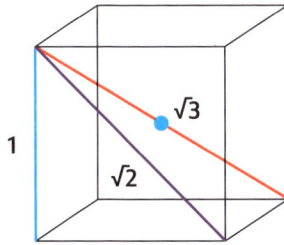

2

The three lengths of a two-dimensional 1:√2 rectangle can be translated into the measurements of a three-dimensional cube.

Note that the √3 line passes through the centre point of the cube, denoted by a blue dot.

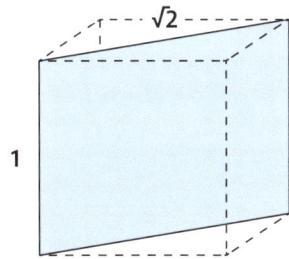

3

The rectangle has the proportion 1:√2. This is because the short edge is the side of a square face with a length of 1, and the long edge is the diagonal of a square face, with a relative length of √2.

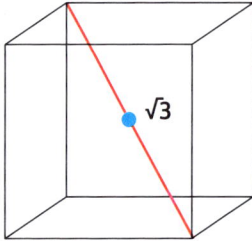

4

The length √3 connects opposite corners of the cube, passing through the centre point.

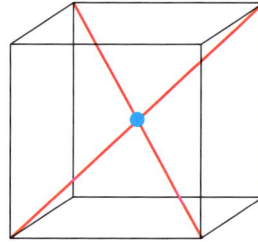

5

A second √3 line can be made to pass through the centre point.

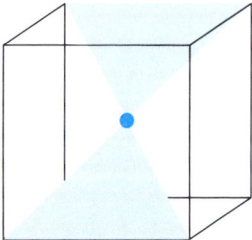

6

These two lines create a flat plane that includes two triangles, as shown.

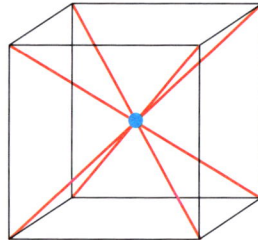

7

A total of four √3 lines can be drawn, creating a pyramid in the bottom half of the cube and an upside-down pyramid in the top half.

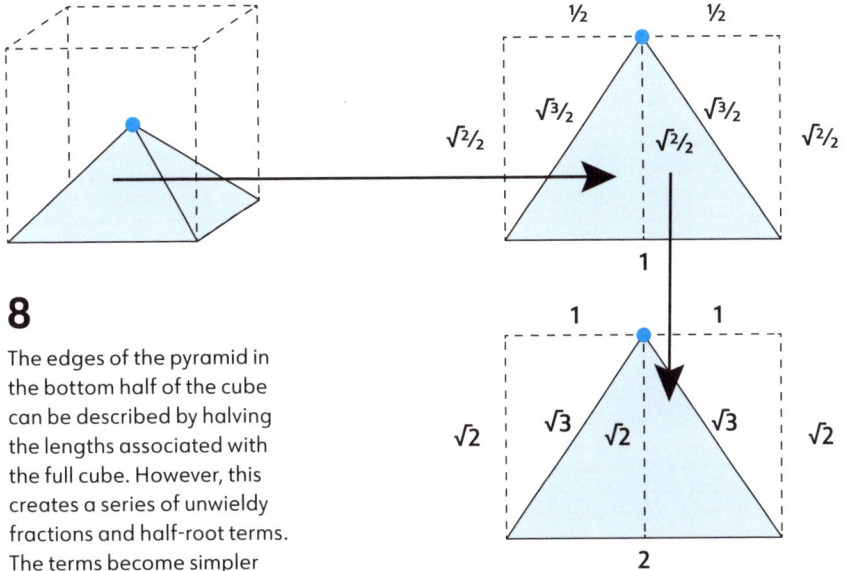

½   ½

$\sqrt{3}/_2$ $\sqrt{3}/_2$

$\sqrt{2}/_2$ $\sqrt{2}/_2$ $\sqrt{2}/_2$

1

1   1

$\sqrt{2}$ $\sqrt{3}$ $\sqrt{2}$ $\sqrt{3}$ $\sqrt{2}$

2

8

The edges of the pyramid in the bottom half of the cube can be described by halving the lengths associated with the full cube. However, this creates a series of unwieldy fractions and half-root terms. The terms become simpler if everything is multiplied by 2, removing the halves and making the side length of the cube 2, not 1. Thus, the base of each triangular face has the length 2 and the sides are $\sqrt{3}$. These proportions are used many times in the remainder of this chapter.

4.11 Sunken Cube

This beautiful three-dimensional form makes maximum use of twelve 1:√2 rectangles, each occupying an edge of a cube and a face common to two pyramids. The folds describe the edges of the pyramids, as explained on pages 106–7. The units are simple to make, but create a sophisticated form.

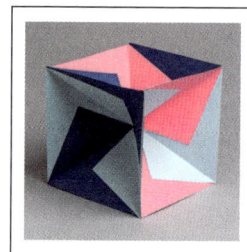

This is the first three-dimensional construction in the book, so you will need to think about how the units connect not only on a flat plane, but also in the third dimension.

The units can be made from any weight of paper or card, but card of approximately 200gsm (54lb) will create a strong result. The base of the triangle should be 12–15cm (5–6in). Use three colours to help you connect the units according to coloured patterns. A structure made from just one colour will be more difficult to assemble.

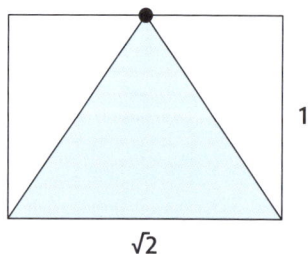

1

$\sqrt{2}$

1

This is the shape of the four triangular faces of each pyramid. Note that the geometry relates strongly to a sheet of 1:√2 paper.

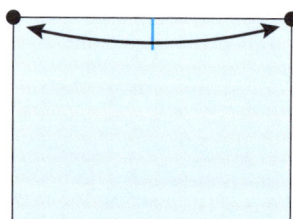

2

Pinch by hand – or measure and mark with a ruler and pencil – the midpoint of a long side.

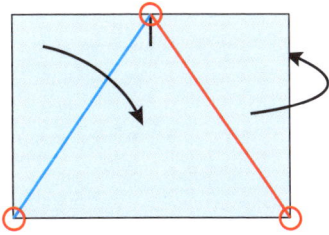

3

Make two folds, as shown, precisely connecting the midpoint with the bottom corners. Note that the left-hand fold is a valley and the right-hand fold a mountain. Make the folds carefully.

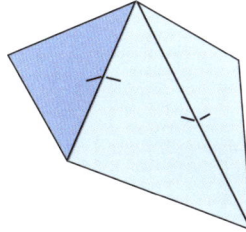

4

This is the result. The top left corner is leaning forwards and the top right corner is leaning backwards.

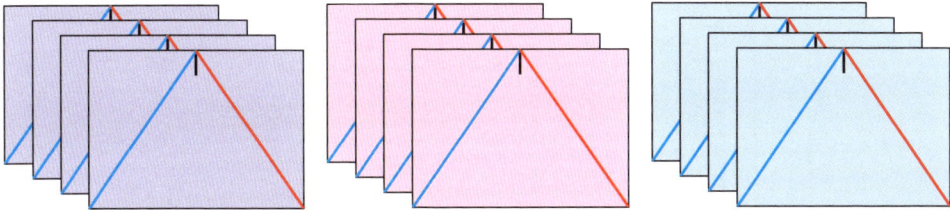

5

Make twelve units: four of one colour, four of another and four of yet another. It is critically important that each unit should have a valley fold on the left and a mountain fold on the right.

If you are using a sheet of A4, a suggestion is to divide it into four 1:√2 (A6) rectangles, to create the four units of one colour.

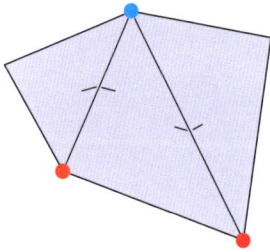

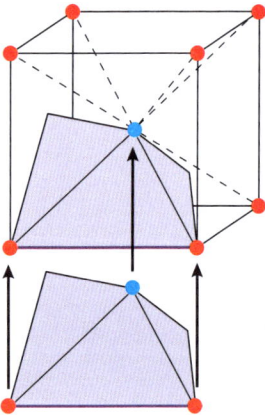

6

The red dot corners on each unit are the corners of the cube. The edge between them is an edge of the cube. The blue dot corner is the centre point of the cube. All twelve blue dots will meet at this central point.

7

The blue dot is the central point of the cube. The dotted lines show where the folds on the units will lie, and the red dots are at the corners of the cube.

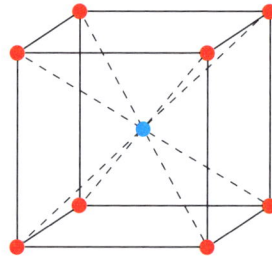

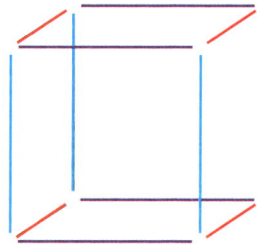

8

This is how one unit will lie inside the cube. Note the alignment of the coloured dots on the unit with the coloured dots on the cube. This alignment will be repeated twelve times around the cube.

9

Step 5 shows the three colours of the units: lilac, pink and blue. The cube shows these three colours along its twelve edges. They are not randomly distributed. Notice how the four edges of each colour are parallel to one another: the lilac edges go from left to right, the pink edges go from front to back, and the blue edges go from top to bottom. The edges of the twelve units will follow this pattern of colour distribution.

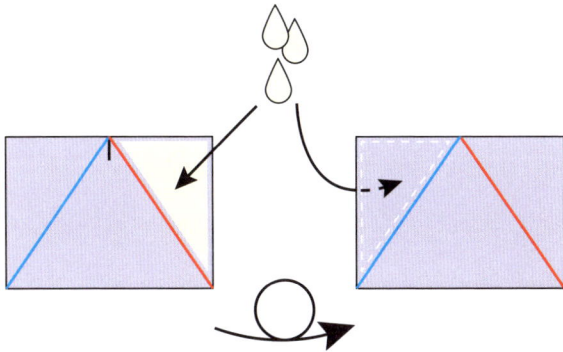

10

Each unit is glued *only* on one triangle, not on both. Be sure to glue the mountain fold side of the triangle. When the unit is turned over, the glued side will be on the reverse, on the left-hand side.

As you make the cube, adding one triangle to another, to another, glue each triangle just before adding it to the others. Don't glue many units at the beginning.

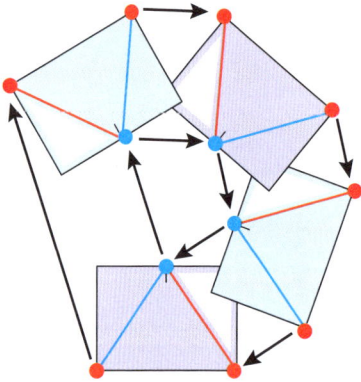

11

Study this glueing pattern carefully. Before glueing anything, hold the units together by hand to understand the structure, perhaps holding them temporarily in place with clips. You will notice that pairs of red dots touch to create a square, and that all the blue dots touch. Note also how the colours of the units are distributed. The drawing may look flat, but the structure is a three-dimensional pyramid. Look at the next step to see what you are trying to achieve. Remember: glue nothing until you have understood what to do.

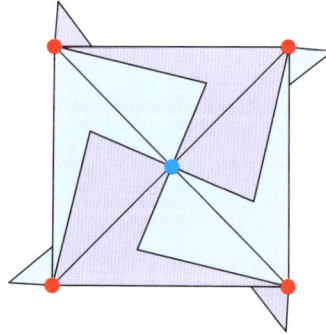

12

This is the result. The blue dot is further from you than the red dots. Note the loose, unglued flaps peeping into view at the back. The structure is a pyramid, with the apex further from you than the base.

Continue to feed in the other eight units, using the loose units at the back as the starting points to create other pyramids. Remember the rules for the blue and red dots, and for the distribution of colours. Eventually, the final sunken cube will be achieved.

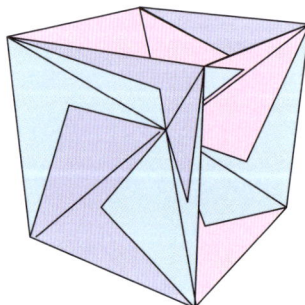

13

Complete. The sunken cube has six pyramids and each pyramid consists of only two colours. The folding is minimal and based exclusively on the relationship between a 1:$\sqrt{2}$ rectangle and a cube.

4.12 Rhombic Dodecahedron

The Rhombic Dodecahedron is neither one of the five very basic Platonic Solids, nor one of the thirteen more complex Archimedean Solids, which together are the foundations of three-dimensional geometry. Nevertheless, it is one of the most beautiful and enigmatic of all solids, consisting of twelve identical rhombic faces. Unsurprisingly, given its place in this chapter, the geometry of the faces is very closely related to that of a 1:$\sqrt{2}$ rectangle.

This section explains a basic method of making the solid, although there are many variations to explore once its basic structure is understood.

Use card of approximately 200gsm (54lb).

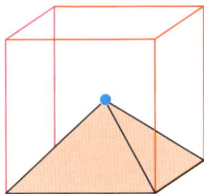

1

The Sunken Cube in the previous section showed how a structure can be built by embedding a pyramid *inside* a cube.

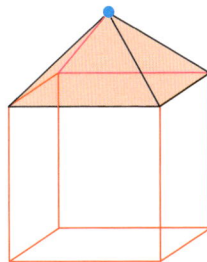

2

By contrast, the Rhombic Dodecahedron is built by placing the same pyramid *outside* the cube, on each of its six faces.

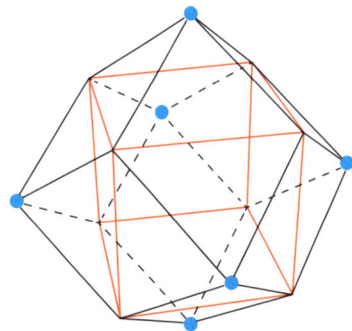

3

Here are the six pyramids, drawn in black. The cube inside is shown in red.

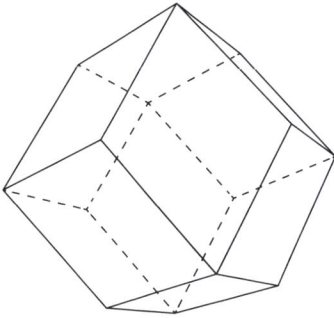

4

Remarkably, when two pyramids meet edge to edge, the two triangular faces form a flat plane. The triangle of one face runs smoothly into the triangle of the other face, so that when the two triangles are combined, they create a flat rhombus. This cannot happen with triangles of any other shape. The illustration shows how the twelve rhombi are distributed. Note how four rhombi meet to form a corner of the Rhombic Dodecahedron when the corners of the rhombi are less than 90°, and three rhombi meet to form a corner of the Rhombic Dodecahedron when the corners of the rhombi are greater than 90°. This 'four and three' structure is a crucial element of its shape.

5

The triangle that forms one half of the rhombic face is exactly the same as the triangle used to form the Sunken Cube. The triangle is made from a 1:√2 rectangle. When two such 1:√2 rectangles are placed together so that the triangles can combine to create a rhombus, that, essentially, is the geometry of the Rhombic Dodecahedron.

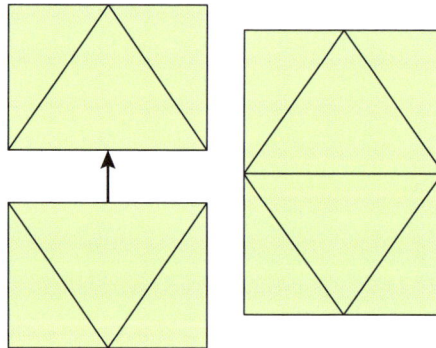

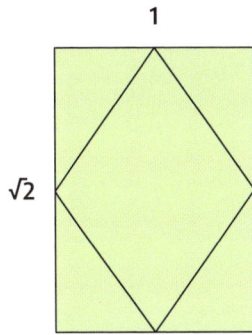

1

$\sqrt{2}$

6

Here is the rhombus. It is formed simply by connecting the midpoints of the four sides of a 1:√2 rectangle.

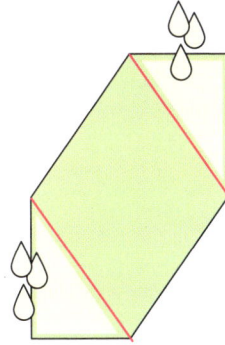

7

To create a unit, cut off an opposing pair of triangles and, at the appropriate time, apply glue to the remaining triangles.

8

Make four units of one colour, four of another and four of yet another. When glue is applied, apply it to the triangles on the mountain side of the folds.

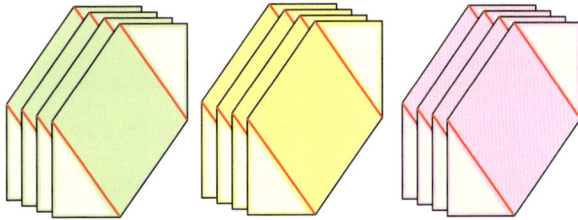

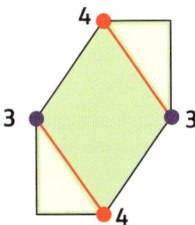

9

4

3   3

4

Here is one of the units. Note that the rhombus has two acute angles (less then 90°) and two obtuse angles (greater than 90°). When you assemble the Rhombic

Dodecahedron, four acute angles meet at a corner and three obtuse angles meet at the other corners. If you follow this pattern, the solid will make itself.

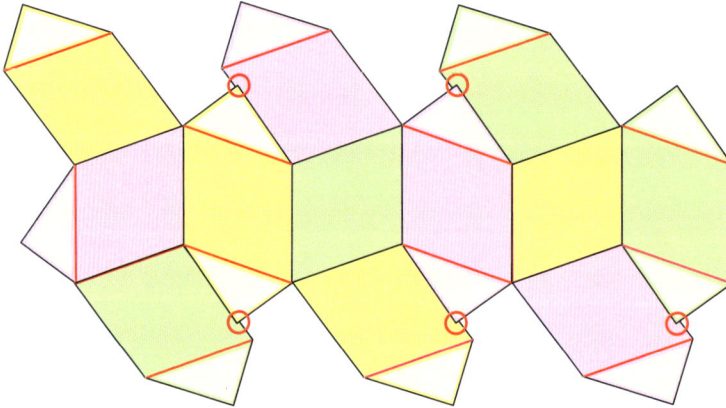

10

There are two ways of assembling the units. One way is shown here. The twelve units are laid out as illustrated and glued together in the colour pattern shown, before glue is applied to the triangles around the perimeter and the solid is glued tight shut in three dimensions. Note that, at the five circled points, the tabs slightly overlap a face. This can be ignored.

The other way to assemble the solid is simply to add one unit at a time to the growing solid, carefully working out (before glue is applied to a new unit) where each unit will go and which colour it must be.

Whichever method you use, pay strict attention to the rules of assembly and the Rhombic Dodecahedron will make itself.

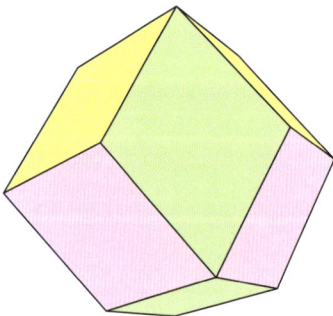

11

Complete. Note the distribution of colours: no two faces of the same colour ever meet edge to edge. A decorative variation is to assemble the form from the same units, but glue the triangles to the outside of a face, not the inside.

From some angles, the Rhombic Dodecahedron appears to have a square silhouette, while from others, it has a hexagonal silhouette. They can be packed together tightly in two-dimensional arrays and three-dimensional grids.

Geometry from A4 Paper

4.13 Half-Pyramids and Quarter-Cubes

We have seen how a specific pyramid, easy to make from a 1:√2 rectangle, can occupy the internal space of a cube to create a Sunken Cube, and can also be placed on the faces of a cube to create a Rhombic Dodecahedron.

Other three-dimensional forms can be made by dissecting the pyramid into smaller forms, then reassembling them in different arrangements, like building with plastic bricks. Here is one suggestion for this dissecting/reassembling technique.

scan for video

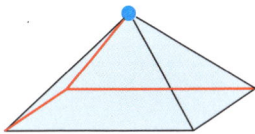

1

Here is the pyramid we used previously. It has four triangular sides and a square base.

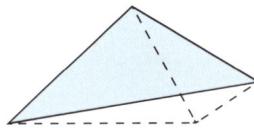

2

The pyramid can be divided into two equal pieces along the diagonal of the square base.

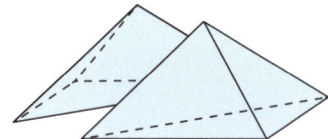

3

If a pyramid made of solid wood were to be cut along this line, the two halves would look like this.

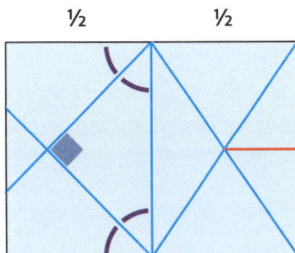

½   ½

4

Here is the method of creating a half-pyramid unit from a 1:√2 rectangle of paper or card.

Fold down the middle of the sheet. Unfold, then, on the right-hand side, make a giant X. Make a short mountain fold, as shown. On the left-hand side, make two creases at 45°. They will intersect at 90°. Note that all the creases are valleys, except for the short mountain.

5

Make a vertical cut through the intersection, as shown. Discard the narrow rectangle.

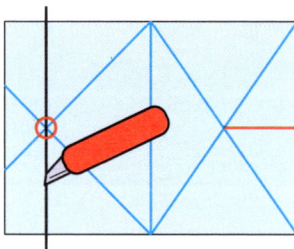

6

Before proceeding, refold all the creases so that they are sharp and strong. The half-pyramid unit can be made by simultaneously collapsing all the creases. Begin collapsing everything ...

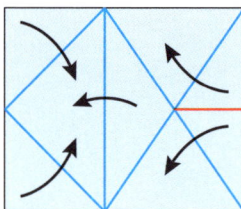

7

... so that the two dots on the right come to touch the single dot on the left. Take careful note of the lettered corners and how they move. Continue to collapse the folds ...

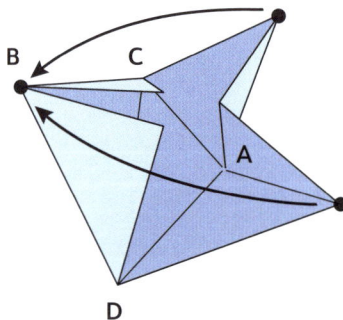

8

... until the mountain fold begins to slide inside, between the triangles at corner B. The three dots will converge at a single point.

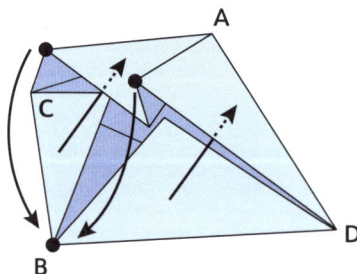

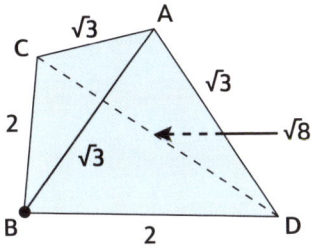

9

This is the final solid. Whatever the weight of paper or card, if it is made carefully and with strong folds, it will hold itself together. However, a little glue placed on the internal flaps will secure it tightly. Note the proportions of the edges.

The solid has three different kinds of triangle. This means that two solids can be joined in three different configurations. Each configuration will have the volume of a complete pyramid, but only one of them will be a pyramid, the others being other forms. Further half-pyramid units can be joined to the initial pair, creating a large number of meta-units with three, four, six, eight, twelve and more units. The possibilities for creating meta-units, then meta-meta-units (joining meta-units together ... and so on) is almost infinite. We will now explore one of these possibilities.

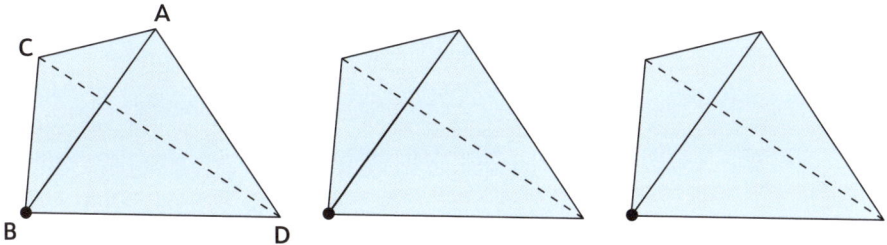

10

Make three half-pyramid units, identical in size, material and colour. Note A, B, C and D.

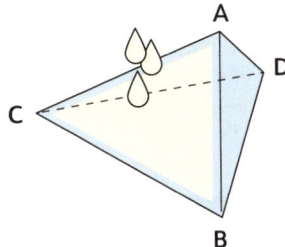

11

Glue is applied to the ABC triangle on each unit. However, before applying glue, look at the next step to understand how the units join together.

12

Note the three A corners and the three B corners. When the units are glued together, all the A corners will touch and all the B corners will touch. The three AB edges will also touch each other. When you have understood this configuration, apply glue as described in the previous step so that the units combine to make a meta-unit.

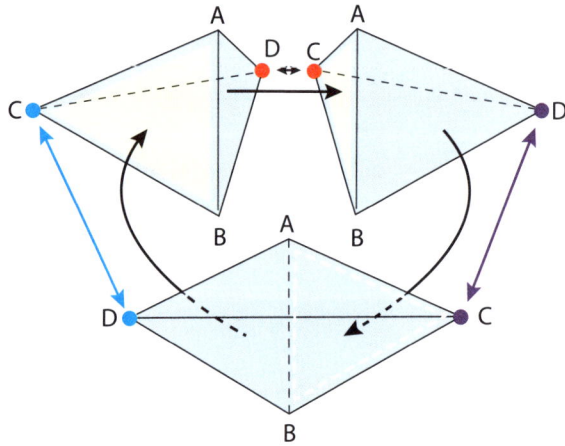

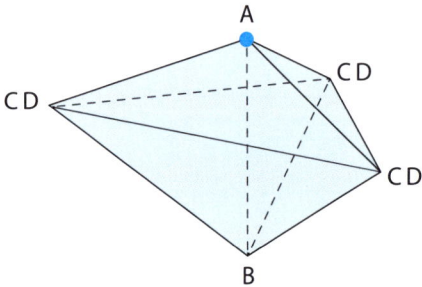

13

This is the final meta-unit. It's a beautiful piece of geometry that the three units each occupy 120° of a 360° circle to lock together so perfectly.

14

There are many interesting ways in which these three-unit meta-units can be joined. Here is one.

Corner B is made of three 90° faces. For this reason, it can occupy a corner of a cube.

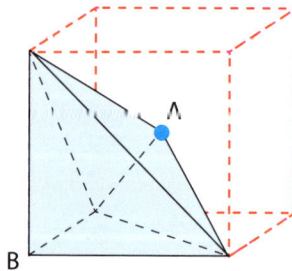

15

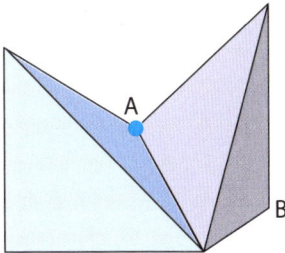

Make another meta-unit and butt it against the first, as shown. The bottom square face of the cube is now complete.

16

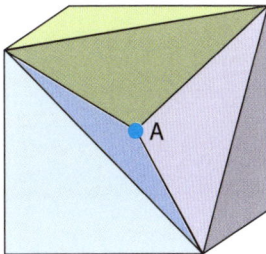

Make a third meta-unit and butt it against the others, as shown.

17

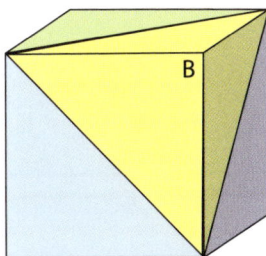

Finally, drop in a fourth meta-unit to create a solid cube. Each meta-unit occupies one quarter of the volume of the cube.

Break the cube apart. How else can the meta-units be arranged? What else can be made if the number of meta-units is increased? There are many beautiful and surprising possibilities to be found.

4.14 One-Sixth-of-a-Cube Unit

The same half-pyramid unit you created in the previous section is used again, but in a different way. Instead of creating meta-units that occupy a quarter of a cube, you will create examples that occupy just one-sixth. Note that, curiously, three of the six units are right-handed (step 8) and three left-handed (step 11).

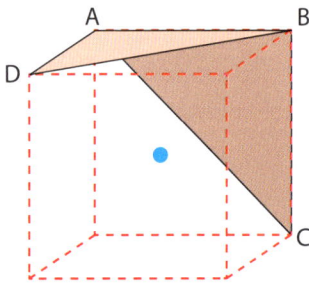

1

For a solid unit to occupy one-sixth of the volume of a cube, it must also occupy one-sixth of the surface. A simple square face is one-sixth of the surface, but we have already explored this possibility with pyramids (see 4.11).

Another way of achieving one-sixth is to divide a face into two triangles and put the triangles on to neighbouring faces, connected by an edge. That is what is shown here. The two half-face triangles are connected along edge AB.

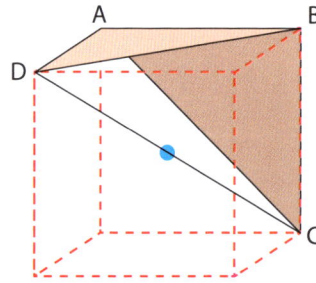

2

To turn a surface unit into a solid unit, we must first create an edge that passes through the centre point of the cube. This is achieved by connecting corner D with corner C.

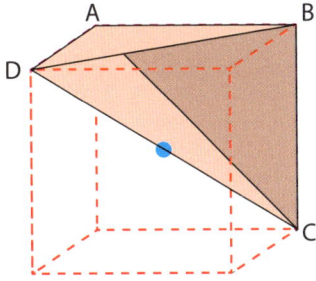

3

Triangles ACD can then be made into a solid face ...

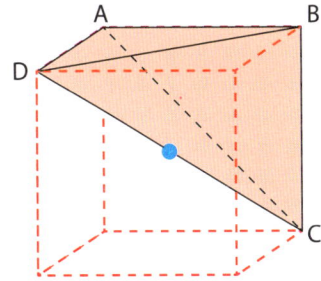

4

... and similarly DBC can be made into a solid face. We now have a solid form with corners at A, B, C and D. Since the original two triangles occupied one-sixth of the surface of a cube, this solid must occupy one-sixth of the volume.

5

This is the geometry of the solid. The relative lengths of the edges (1, √2 and √3) will be familiar.

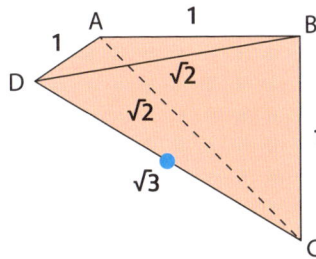

6

The solid can be made by combining two half-pyramid units, made in the previous section. The way in which the units join is subtle and must be understood before any glue is applied. The unit on the left has its 90° corner flat on the ground. The unit on the right has its 90° corner standing upright at bottom right.

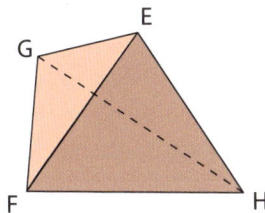

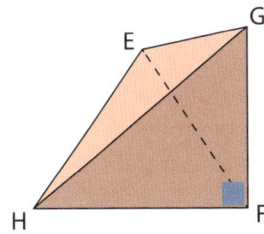

4.14   One-Sixth-of-a-Cube Unit

7

Bring the solids together, as shown, so that corner E touches corner E, and H touches F. Only when you understand the alignment should glue be applied.

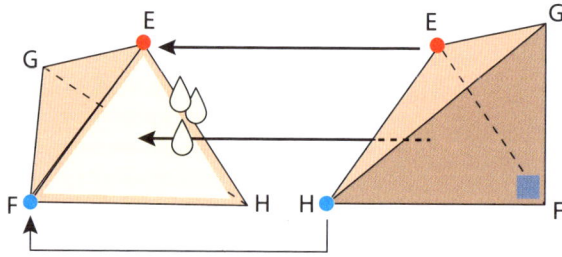

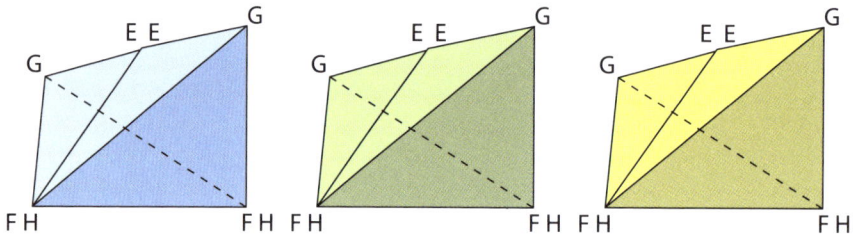

8

This is the assembled two-unit meta-unit. It looks very different from the half-pyramid unit with which we began our journey.

Now make two copies of this meta-unit, being careful to assemble the two units in the same way and not as a mirror image of the blue meta-unit.

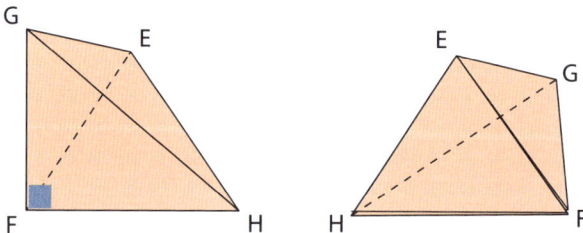

9

We must now create three meta-units that are mirror images of the three just made.

Look very carefully at the illustration and, before using glue, check that what you are making is the mirror image of the three you just made. This can take a few moments to understand, so be patient.

10

This is how the three new meta-units are created.

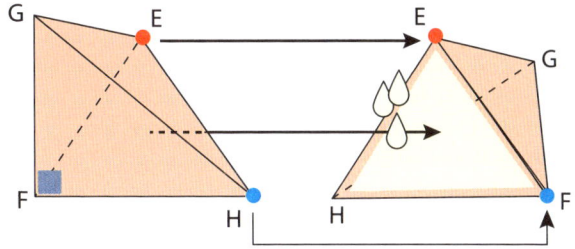

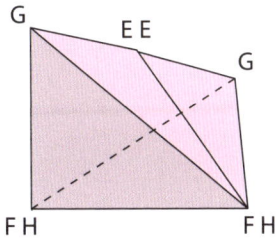

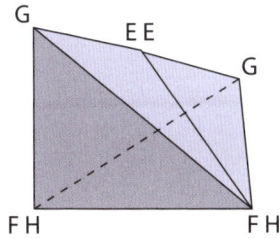

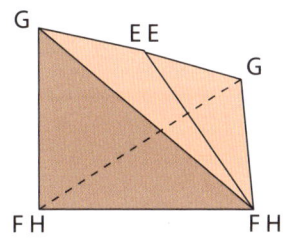

11

Create three new meta-units. Check that they are the mirror image of the first three meta-units (step 8).

12

This shows the relationship of the meta-unit to a cube. Note that the two triangles on the surface of the cube are in the same positions as the triangles in step 1.

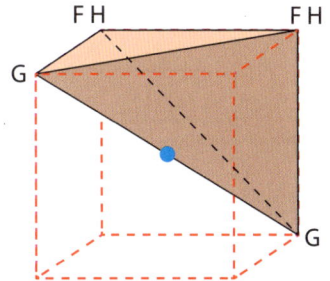

13

Make a second meta-unit and position it as shown. This unit is the mirror of the first.

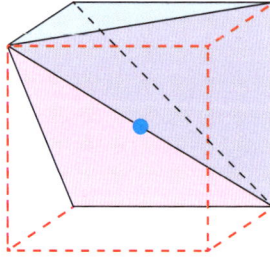

14

Add a third unit. This unit is the mirror of the second.

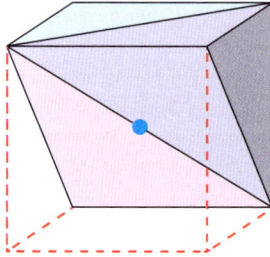

15

Add a fourth unit. This unit is the mirror of the third.

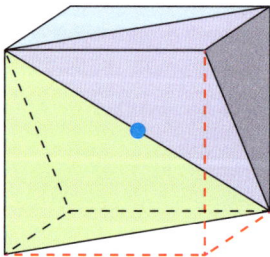

16

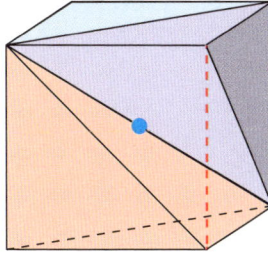

Add a fifth unit. This unit is the mirror of the fourth.

17

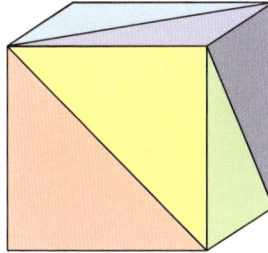

Finally, add a sixth unit. This unit will be the mirror of the fifth and first units. The cube is complete.

By joining half-pyramids in a variety of ways, you can create a great many other forms.

A pyramid can also be divided into a quarter unit, then those units rearranged and joined in different configurations to create meta-units, which in turn can combine to create many different larger solids.

How else can a cube be divided into an equal number of solids? Do we have to begin from a pyramid? Absolutely not! Think about how to divide the surface of a cube into halves, thirds, quarters and other divisions, then think about how those surfaces can be made solid.

Since the basic lengths in a cube (1, $\sqrt{2}$ and $\sqrt{3}$) are also those of 1:$\sqrt{2}$ paper, with a little creative thinking it is generally easy to make a solid unit from such a sheet, although a little trimming or glueing may sometimes be necessary.

This is an immensely creative topic to explore. With only a little experimentation, you will soon be creating forms that no one has seen before.

⑤

Basic Solids

5
Basic Solids

There are five so-called Platonic Solids, that is, solids made from only one kind of regular polygon, whose corners are all equidistant from a central point. They are the simplest of all solids and the basis of all three-dimensional geometric forms.

The methods of making the Platonic Solids shown in this chapter have been chosen for their technical diversity. The tetrahedron is pure origami (folded, no glue), the cube is pure modular origami (folded, no glue), the octahedron is cut-fold paper engineering (folded and glued), the dodecahedron is a nearly accurate approximation of dodecahedral geometry, and the icosahedron is a one-fold engineered unit (little folding, much glueing). The chapter also shows two ways of making a number of pyramids.

As a collection of techniques, they cover the different ways in which any solid can be made.

5.1   Tetrahedron

The tetrahedron is the simplest of all the solids, having just four equilateral triangles as faces, plus four corners and six edges. This method uses an A4 sheet and is perhaps the simplest and most direct way of making a tetrahedron. Note how it uses the method of making a 60° angle shown in 1.3.

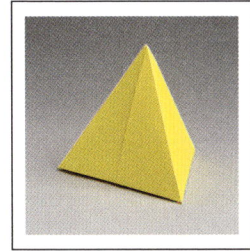

scan for video

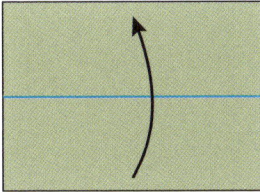

1

Use a sheet of A4 paper, or paper proportioned 1:√2. Fold one long side to the opposite long side.

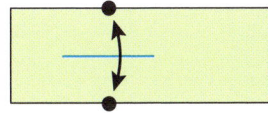

2

Fold across the middle of the top layer only. Note that the fold need not stretch across the full width of the paper.

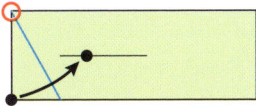

3

Check that the fold you made in step 1 runs across the bottom of the paper, then fold dot to dot, as shown.

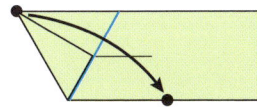

4

Again, fold dot to dot.

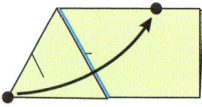

5

And again, dot to dot.

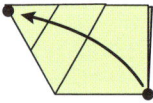

6

And one final time, dot to dot. Note that the dot at the top left is not exactly at the corner of the paper. If you are using A4 or 1:√2 paper, the dot will be a little way down the sloping edge.

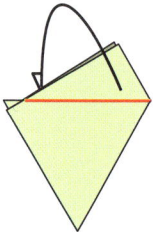

7

Fold the excess triangle at the top to the back.

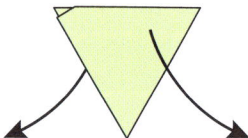

8

Unfold everything back to step 2.

9

Note that you have created a chain of equilateral triangles across the paper, beginning by creating a 60° angle in step 3. Strengthen all the folds, one at a time.

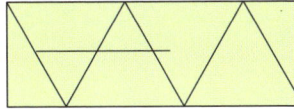

10

All the folds now move at once. Tuck edge BC between the layers at AD.

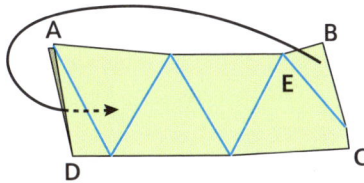

11

Push BC inside the layers until E touches A. When this happens, the paper will have taken the shape of a tetrahedron. Strengthen all the folds.

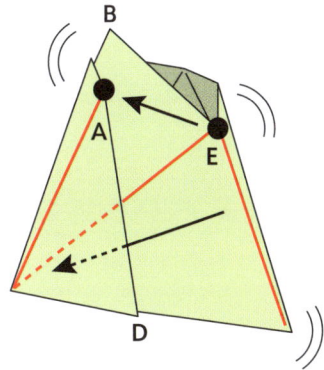

12

Complete. This is a surprisingly simple way of creating a stable tetrahedron. There are a great many ways of making the same form by collapsing a complex grid of equilateral triangles, but none can match the method shown here for speed and lack of fuss.

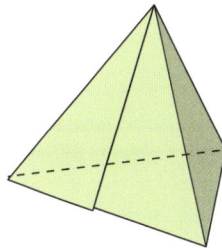

5.2 Cube

There are countless ways of making cubes by collapsing a pre-creased grid of squares, by folding strips of paper, and by combining any number of folded units. Some are plain and simple, others highly decorative. The method shown here – the so-called Jackson Cube – was created by the author in the early 1980s and is arguably the simplest way of creating a clean cube.

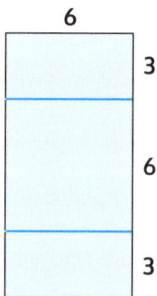

1

Use light card of approximately 200gsm (54lb). Cut a rectangle 12 × 6cm (5 × 2½in) and accurately score folds 3cm (1¼in) from each end. Note that the rectangle is proportioned exactly 2:1. The unit may be made larger, but the card will need to be heavier as the scale increases, or the result will be a weak lock.

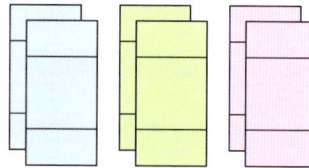

2

Make six units. To aid assembly, it helps to have three pairs of colours. Be sure to make them accurately.

3

The two creases should each be folded to 90°.

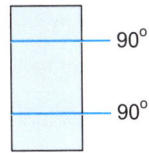

90°

90°

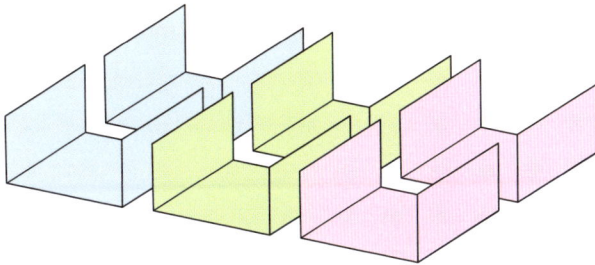

4

Fold each of the six units in the same way.

5

There are many unstable ways of assembling the six units into the shape of a cube, but only one that will hold together strongly. Finding that one way can be a challenging puzzle. Here is the method.

Each unit consists of a central square and, at opposite ends, a pair of tabs. Lay one unit on its back. Lock into it two units of another colour, such that their tabs lie on top of the first tab, as shown. Note carefully the position of the tabs on all three units.

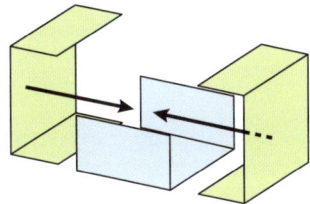

6

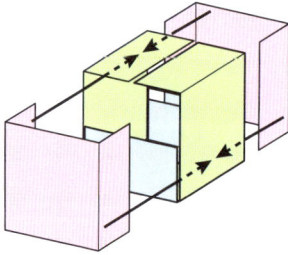

Introduce the two units of the third colour. Note that their tabs are positioned to the left and right of the units.

7

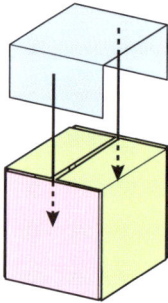

Finally, slot in the remaining unit on top.

8

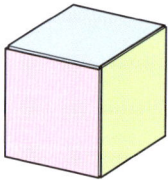

Made carefully, the cube will be extremely strong. Note that each of the twelve edges is both the raw edge of one unit and the folded edge of another, so it is impossible to insert a finger into the cube. Note also that no tab ever touches another. The structure is very symmetrical, and it is this symmetry that gives the cube strength.

Now give the pieces to a friend and challenge them to assemble the cube.

5.3   Octahedron

The octahedron has eight equilateral triangles, six corners and twelve edges. The method of constructing this version is a hybrid between origami and paper engineering, since the grids of equilateral triangles are created by folding and the two units are assembled with glue. As with the tetrahedron, the primary task is to create a grid of equilateral triangles from which the shape of each unit can be cut and glued.

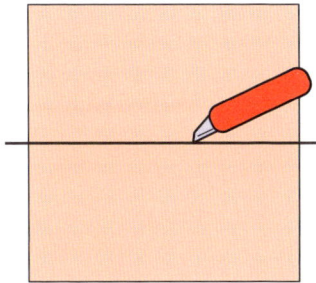

1

Begin with a square of paper or light card, approximately 20–25cm (8–10in) along each side. Cut the square in half.

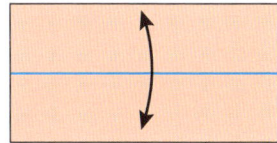

2

Fold one of the halves in half. Unfold.

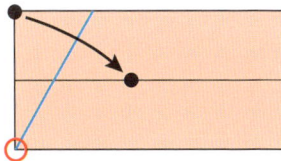

3

Fold dot to dot, as shown, to create an angle of 60°.

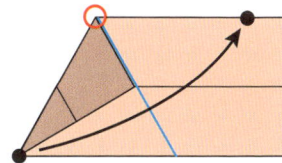

4

Again, fold dot to dot, as shown.

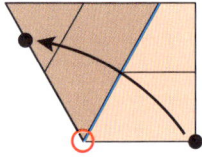

5

And again, dot to dot.

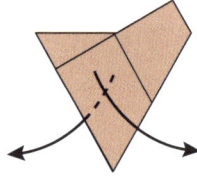

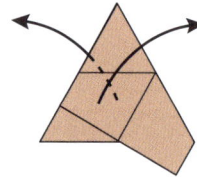

6

Unfold everything.

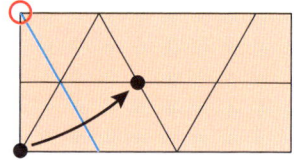

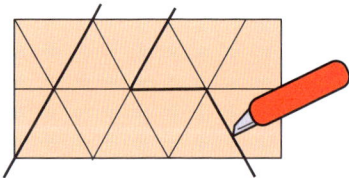

7

Repeat steps 3–5, this time folding dot to dot using the bottom corner.

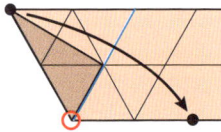

8

Fold dot to dot, as shown.

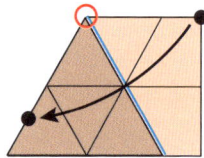

9

And again, dot to dot.

10

Unfold everything.

11

Make two cuts, as shown, one straight, the other a zigzag. After the excess triangles have been removed, seven remain in the middle of the sheet.

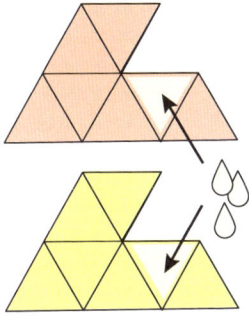

12

Make a copy. You may use the other half of the square, but for clarity, a second colour will be shown here.

Apply glue to one triangle on each unit, as shown.

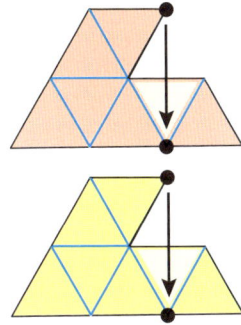

13

Fold dot to dot so that the units become three-dimensional. Check the next step to see how the units should look.

14

The octahedron is made by placing one unit on top of another.

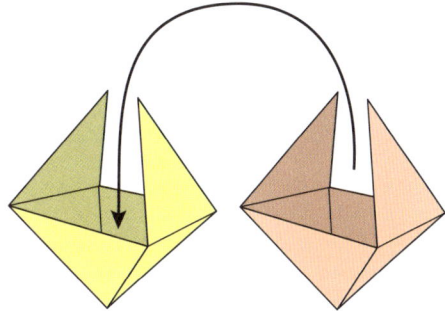

15

Complete. The four loose triangles can either be tucked inside the opposing half or left outside. Both configurations are possible, and each will create its own distinct colour pattern across the eight triangles on the surface. Use glue to hold the triangles in place, inside or outside.

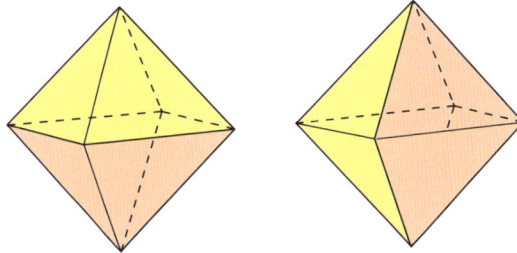

5.4   Dodecahedron

A dodecahedron has twelve pentagonal faces, twenty corners and thirty edges. The three preceding projects in this chapter have all created clean, solid polyhedrons. This project is different because it is made from thirty edge units that combine to create a decorative surface.

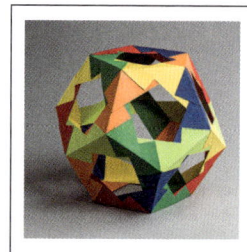

The unit is adapted from an origami unit created by Robert E. Neale (US) in the 1960s.

scan for video

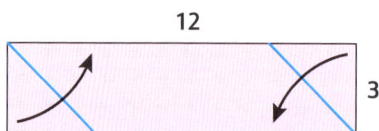

1

Use a 4 × 1 rectangle of light card, perhaps 12 × 3cm (4 × 1in). Whatever the size, the proportion of the unit must be 4:1.

On the left-hand side, fold the bottom corner upwards at 45°. On the right-hand side, fold the top corner downwards at 45°.

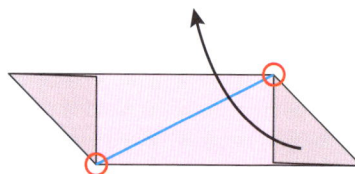

2

Fold exactly between the obtuse corners, as shown. This must be done accurately, so perhaps use a knife and ruler to help you score the fold.

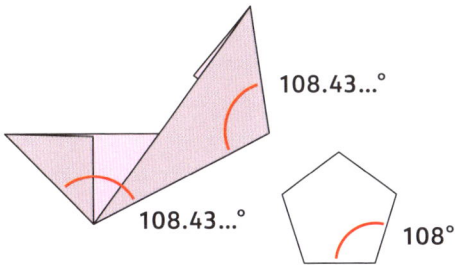

3

The resulting angles will be 108.43...°, very close to the interior angle of 108° in a pentagon. Since a dodecahedron is made from twelve pentagonal faces, this angle is critical. The error of less than 0.5° is nominal and will not be a factor when you assemble the units, provided the construction is accurate.

4

Make thirty identical units, one for each edge of the dodecahedron. They can all be the same colour, or you can make five units in each of six different colours. Be sure that no unit is made as a mirror image of the others, and that they are all made accurately.

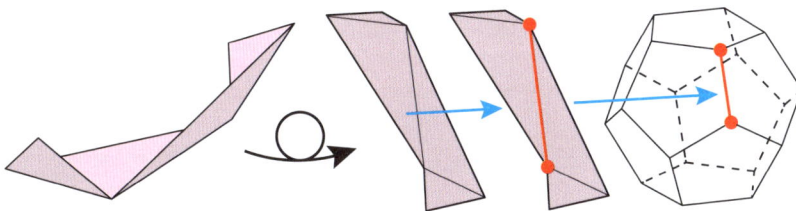

5

This is a typical unit. Turn it over so that you can see three mountain folds.

6

It is important to understand how the structure of a unit relates to the structure of the dodecahedron. The long central fold is an edge of the dodecahedron. The two red dots show two corners. Each unit thus contributes one edge to the structure, and part of two adjacent corners.

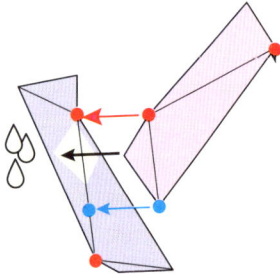

7

Three units combine to create each corner. The red dots on two units will lie on top of each other, such that the short fold is on top of the long fold …

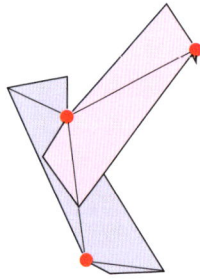

8

… like this. Before applying glue, be sure you understand how the two units lie in relation to each other.

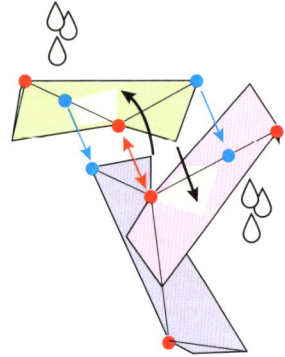

9

A third unit is introduced to complete the corner. As before, apply glue only once you are sure you understand how the three units lie in relation to one another.

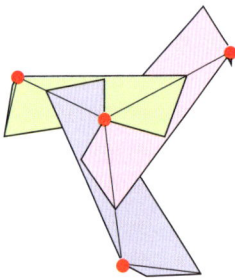

10

This is the result. Note that the three units are arranged symmetrically around the corner, that short folds always lie on top of long folds, and that accuracy is important. Take your time.

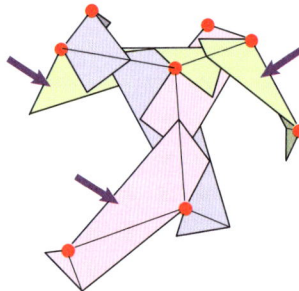

11

Since each unit forms part of two corners, each of the three units in the assembled corner will also be part of other corners. Here, three more units have been added, indicated by the arrows.

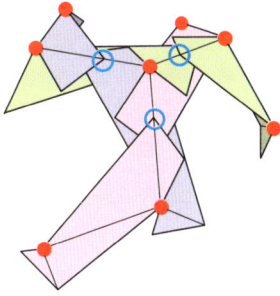

12

In order to distribute the six colours correctly, there is a rule that states where they should go. First, note that the assembled corner (step 10) uses units coloured lilac, green and pink. The new units added now are the same colours. Inside the circles, the new lilac unit is overlapping the old lilac, the new green is overlapping the old green, and the new pink is overlapping the old pink, completely hiding the long folds underneath. This overlapping of like colours happens on every edge. If followed precisely, the six colours will be distributed symmetrically around the dodecahedron.

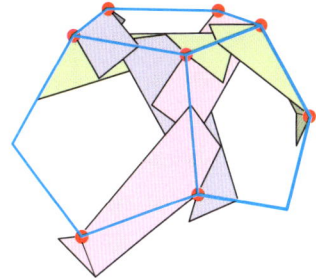

13

The blue pentagons show how the growing assembly of units relates to the shape of the dodecahedron.

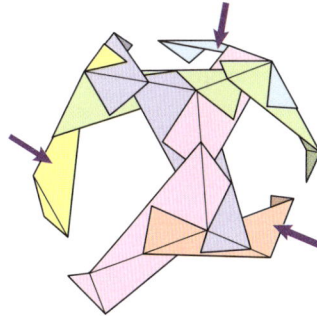

14

You will notice that like colours overlap, meaning each colour will form a continuous band of overlapping units around the dodecahedron. These six bands circulate in different directions and combine to create the dodecahedron. This graphic shows how the six units of one colour are distributed. The pattern is the same for each colour.

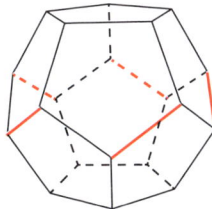

15

Three more units have been added, using the new colours. The other units can be added, one at a time, always following the rule that like colours overlap. It's a challenge to do this correctly, but the result is spectacular.

5.5   Icosahedron

An icosahedron has twenty equilateral triangle faces, twelve corners and thirty edges. The version shown here is similar to the dodecahedron unit, in that each unit is an edge. The unit is very simple to make – it has just one fold – but when assembled, the resulting icosahedron is spectacular.

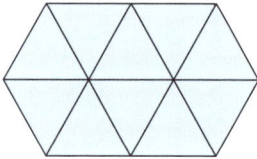

1

This is the geometric shape of the units, a collection of ten small equilateral triangles. Make them from light card so that the fold across the middle is approximately 10cm (4in) long.

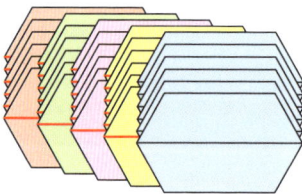

2

This is a single unit, without the ten equilateral triangles.

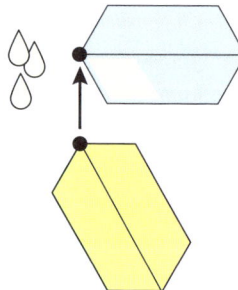

3

Make thirty units. They can all be the same colour, but if you want to use different colours, use five colours and make six units of each colour.

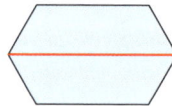

4

The fold across the middle of each unit is an edge of the icosahedron. Two units connect as shown, dot to dot. Use glue.

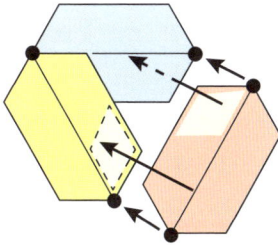

5

A third unit will interlock with the other two, as shown. Be careful to keep the corners tidy and accurate.

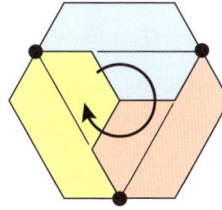

6

A triangular face is complete. Note the symmetry and the clockwise rotation of the surface. This clockwise rotation will be repeated on every face.

7

The triangular face completed in the previous step is at the bottom left of the illustration. Continue to add more units in the colour pattern shown.

Note that there are four completed triangles around the centre point. Close the gap so that five complete triangular faces meet at the central corner. When this is done, the central corner will rise to create a five-sided pyramid.

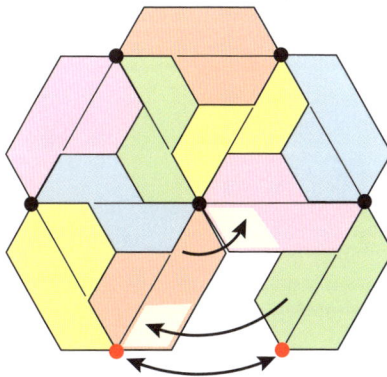

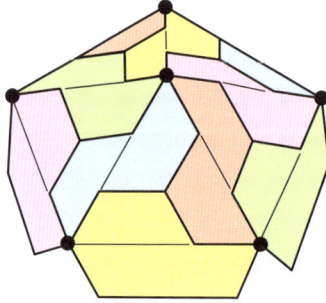

8

This is the completed five-sided pyramid. Note that to make an icosahedron, five triangular faces and five edges must meet at every corner.

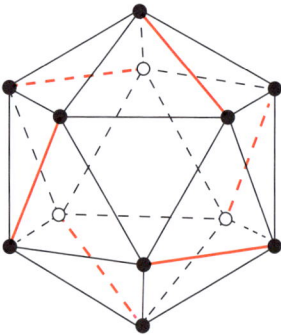

9

How are the colours distributed symmetrically? In truth, the pattern is a little complicated, but there is just one rule.

Identify an edge (shown here in red). Look at the corner of the triangular face not touched by that edge. Five edges touch that corner. The edge with the same colour as the edge you first identified will be the middle edge of the five. It is perpendicular to the first identified edge. Continue this pattern around the surface.

This rule will allow six edges of the same colour to be distributed symmetrically around the icosahedron.

5.6 Pyramids

After the five Platonic Solids, pyramids are perhaps the most common three-dimensional forms to make. There are two basic ways of making them, depending on their pitch (how spiky or squat they are) and how many sides they have. Both methods are explained here.

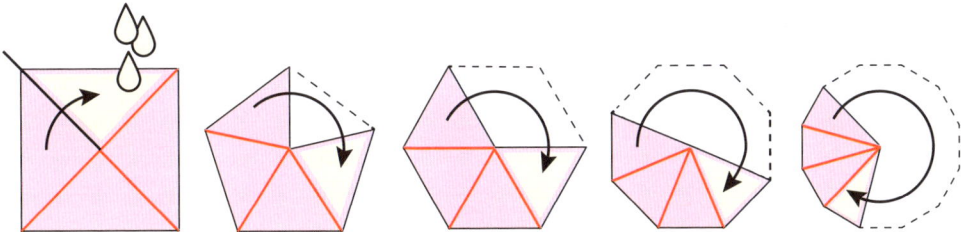

1

The Pitch

Here are five ways of making a three-sided pyramid. Each is made from a regular polygon with (from left to right) four, five, six, eight and twelve sides. Each example has three intact triangular faces and one extra face that can be overlapped with one of the intact triangles to close the pyramid.

Note that the angle at the apex of a pyramid decreases as the number of the sides of the polygon increases. Thus, the squattest pyramid is made from a square, whereas the spikiest is made from a dodecagon (twelve-sided polygon).

Making pyramids from polygons is the easiest way of constructing them. If you need something more precise, the examples shown here will give you an approximate guide, after which you can calculate the angles more exactly.

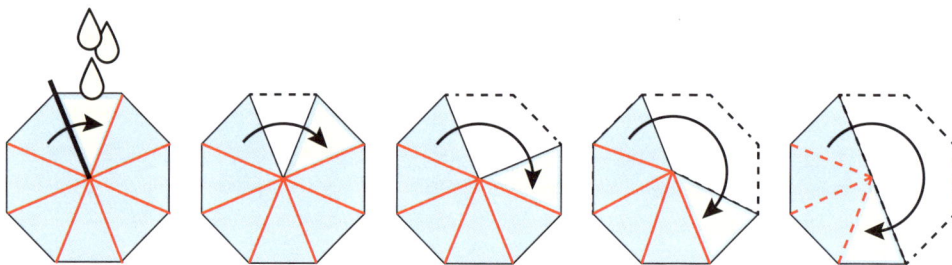

2
The Number of Sides

Here are five ways in which an octagon can be used to create pyramids with three, four, five, six and seven sides. The fewer sides the pyramids have, the greater the pitch, despite the fact that the faces are always identical.

The same variations can be made using other polygons.

The more sides a polygon has, the more pyramids can be made from it.

Once you understand the principles of making pyramids by controlling the pitch and the number of sides, you can use them in collaboration to create a vast number of different examples.

4.11 Sunken cube

4.12 Rhombic dodecahedron

4.13 Cube from four meta-units

4.14 Cube from six meta-units

5.1 Tetrahedron

5.2 Cube

5.3 Octahedron

5.4 Dodecahedron

5.5 Icosahedron

5.6 Pyramids

5.6.1 Pyramids

Reflections

Did you make anything? If you did, what was your experience like? Which examples were the most useful, or the most surprising, or the most beautiful … and why?

Or did you skim through the pages, understanding and perhaps admiring the examples, but not breaking out the paper? Either way, the book should have demonstrated that folded geometry has a place in your toolkit of making techniques, whatever your specialism as a designer or maker.

The word 'geometry' comes from the Greek *geo*, meaning Earth, and *metry* – or *metre* – meaning measure, so geometry has its origins in measuring the Earth. It is a physical, real-world subject, developed to help us build and make with accuracy, safety and strength. Although the ancients used straight edges, compasses and plumb lines to help them construct objects large and small, folding paper offers a tabletop alternative for the contemporary designer-maker. It helps to bring us back to our roots as hands-on workers, not wholly dependent on software, renderings and AI for our output. Essentially, folding paper in this way puts us back in control of our work.

With robots and AI increasingly taking over production, one scenario for our mid-term future is that it will be those with hand skills who will be the most employable and the most sought-after. Is that you? It may seem that we are increasingly slaves to technology, but perhaps the opposite is more true: that, as a counterpoint, we need – more than ever – people who can build and make with their hands. Folded paper geometry fits this mode of production. Students applying for design courses seem generally to have fewer hand skills than in previous generations, which in turn has created a generation of designers who have less awareness of materials and three-dimensionality outside their two-dimensional computer screen. This is to be regretted.

So, in your own way, take back control of your work. Make more, beginning with the basics. Folded paper geometry will help you achieve this.

Acknowledgements

I've been a professional origami artist and teacher/lecturer for more than forty years. During this time, I've had countless opportunities to learn folded geometry from brilliant people across the wide world of origami. To thank everyone from whom I've learned something would be to thank scores, if not hundreds of people, so I offer a profound and collective thank-you to my inspirational friends and colleagues everywhere.

I must also thank the many colleges of design across a dozen or so countries that have given me abundant opportunities to run paper manipulation courses for their students. Teaching has given me a strong belief that folded geometry is important. It has also given me ample opportunities to understand what to teach and how to teach it.

Finally, working with my wife, the origami artist Miri Golan, on her multi-award-winning Origametria programme – which uses origami to teach curriculum geometry to primary-school students – helped me to divide the pretty from the practical and the essential from the non-essential so that, in my own mind at least, I came to understand what should be included in a book about folded geometry. Thank you, Miri.